Metabolic Control of Eating, Energy Expenditure and the Bioenergetics of Obesity

Volume Editor

Artemis P. Simopoulos
The Center for Genetics, Nutrition and Health,
American Association for World Health, Washington, D.C.

33 figures and 27 tables, 1992

KARGER

Basel · Freiburg · Paris · London · New York · New Delhi · Bangkok · Singapore · Tokyo · Sydney

World Review of Nutrition and Dietetics

Library of Congress Cataloging-in-Publication Data
Metabolic control of eating, energy expenditure, and the bioenergetics of obesity /
volume editor, Artemis P. Simopoulos.
(World review of nutrition and dietetics; vol. 70)
Includes bibliographical references and index.
1. Energy Metabolism. 2. Appetite. 3. Obesity. 4. Bioenergetics. 5. Energy Metabolism.
I. Simopoulos, Artemis P., 1933–   . II. Series.
[DNLM: 1. Biological Transport. 2. Feeding Behavior – physiology.
3. Obesity – metabolism.]
ISBN 3–8055–5595–4 (alk. paper)

Bibliographic Indices
This publication is listed in bibliographic services, including Current Contents® and Index Medicus.

Metabolic Control of Eating, Energy Expenditure and the
Bioenergetics of Obesity

World Review of Nutrition and Dietetics

Vol. 70

KARGER

Basel · Freiburg · Paris · London · New York · New Delhi · Bangkok · Singapore · Tokyo · Sydney

Contents

Contents

The Bioenergetics of Obesity Syndrome

Contents

Preface

Interest in the control of body weight either through diet or exercise or both has led to research of the mechanisms involved in the metabolic control of eating and in energy expenditure. Obesity is the most common metabolic disorder with nutritional components, that continues to be on the increase in the developed world. Its incidence is also on the increase in the developing world where it is found alongside malnutrition due to undernutrition.

This volume consists of three papers. It begins with the paper on the 'Metabolic Control of Eating'. This paper includes sections on eating and energy balance; meal-taking behavior; effects of circulating fuels (glucose, nonesterified fatty acids and other fuels) on eating; and the various controls that exist such as control of eating by hepatic metabolism, by pancreatic hormones, metabolic sensors in the brain, and metabolic cues and taste perception. Interactions among the various factors of chemo- and mechanoreceptors in the mouth, stomach and intestine lead to the concomitant release of gastrointestinal and pancreatic hormones. Each of these factors appears to be an important modulator of the satiating potency of other factors, thus ensuring successful adaptation of eating behavior to physiological needs. The author concludes that 'separate receptors monitor glucose utilization and the oxidation of other fuels, hence implicating that the integration of the separate signals occurs on the neural level ... although some integration may occur on a biochemical level ... despite the considerable knowledge accumulated during recent years, our understanding of the metabolic control of eating is certainly still incomplete.'

The second paper 'Energy Expenditure and Fuel Selection in Biological Systems: The Theory and Practice of Calculations Based on Indirect Calorimetry and Tracer Methods' is an authoritative and comprehensive review on calculating energy expenditure and fuel selection, with particular reference to composition of foods. The subject is of central importance to a variety of nutritional studies. The authors illustrate the origins of the theory and have developed procedures which can be readily applied to various situations. In the introduction the authors state 'The reliability of indirect calorimetry for estimating energy expenditure and fuel selection is strongly dependent on the use of appropriate fuel coefficients. The fuel coefficients usually adopted are general ones, but they vary from one fuel to another. Recently there has been an increasing use of nutritional regimens of unusual composition, especially with artificially compounded feeds that are administered enterally or parenterally in hospitals, or of native foods in non-industrial developing regions. Therefore, it has been necessary to reconsider the principles of indirect calorimetry, particularly the use of fuel coefficients in different circumstances. Also the increasing use of tracer techniques, which involve estimating CO_2 production alone, has made it necessary to consider in detail the variability of energy equivalent of CO_2 and its predictability in different circumstances.' The paper begins with a discussion on the principles of indirect calorimetry followed by a section on establishing calorimetric factors (fat, carbohydrate, protein). The third section is on derivation of equations used to calculate energy expenditure and fuel selection. The fourth section is on estimating the proportion of energy derived from the oxidation of different fluids. The next section is on the estimation of ATP gain which is followed by a section on the consequences of using different coefficients for individual fuels to calculate energy expenditure and fuel selection. The final section is on the assessment of the errors. The principles of tracer techniques for measuring energy expenditure in free-living subjects, such as the double-labelled water method and the labelled bicarbonate method, are included in this review with emphasis on the errors arising from the application of inappropriate energy equivalents of carbon dioxide.

Obesity represents mostly the chronic effects of imbalance between energy intake and energy expenditure. How this imbalance translates into the obese state and the mechanisms involved has occupied the interest of many scientists in various fields. The next paper on 'The Bioenergetics of Obesity Syndrome' is a major review of research over a 10-year period from 1981 to 1991. Obesity has many causes – genetic, hormonal, dietary, and sedentary life styles – therefore, in this paper it is referred to as obesity syndrome rather than obesity. The paper presents in an orderly fashion the various causes and consequences of obesity; the mechanisms of cellular energy transduction with extensive discussions about the mitochondrion's functions and disfunctions in both malnutrition and overnutrition; in the genetically obese mouse, and the uncoupling of mitochondria by endogenous and exogenous substances. The roles of defective energy metabolism in the etiology of obesity is considered in terms of defective thermogenesis; changes in resting basal metabolic rate in dietary obesity; Na^+K^+-ATPase status and changes in oligomycin-sensitive ATPase in dietary obesity; the energy implications of cyclical obesity in hibernators; changes in key glycolytic enzymes in obesity; and the limitations of direct and indirect calorimetry in the assessment of bioenergetics in obesity. The section on mitochondrial status and related energy-linked functions of dietary obese animals consists of reviews and summaries of the research in this area including phospholipids and cholesterol changes in the mitochondrial membranes of dietary obese animals; deviation from vectorial chemosmosis and changes in oxidative phosphorylation characteristic of obese animals; brown adipose tissue mitochondria and thermogenesis in obesity; and respiratory impairment and low-work capacity in obese subjects. In the last part of this section the question is posed, 'Does a dietary obese animal rely more on glycolysis?' The author answers his own question in the affirmative, since the mitochondria isolated from the obese animals do not function normally.

The concept of thermogenic defect in subjects who are obese is a very intriguing one and a number of studies continue to contribute to its understanding. Of interest is the fact that those obese subjects in whom a thermogenic defect was observed, were selected on the basis of family history of obesity, suggesting that the subjects were similar genetically. In other studies, not selected on the basis of family history, a thermogenic defect was not shown. These conflicting findings emphasize the need to use approaches that take into consideration the methods of genetic epidemiology in carrying out such studies.

The three papers in this volume constitute the most up-to-date reviews on the metabolic control of eating, energy expenditure and the bioenergetics of obesity. These are three subjects which are of major importance to the study of obesity or obesities. The student and the researcher are given an opportunity to have in just one single volume the most authoritative reviews. This volume should be of interest to physiologists, biochemists, nutritionists, physicians with particular interest in obesity and clinical nutrition, dietitians, and researchers in the fields of pharmacology, exercise physiology, membrane biochemistry, endocrinology, metabolism and genetics.

Artemis P. Simopoulos, MD

Simopoulos AP (ed): Metabolic Control of Eating, Energy Expenditure and the Bioenergetics
of Obesity. World Rev Nutr Diet. Basel, Karger, 1992, vol 70, pp 1–67

Metabolic Control of Eating[1]

W. Langhans[a]*, E. Scharrer*[b]

[a] Institute of Animal Sciences, Swiss Federal Institute of Technology and
[b] Institute of Veterinary Physiology, University of Zürich, Switzerland

Contents

[1] The support of the Swiss National Science Foundation (grants No. 3.944-085 and
32-027854) is gratefully acknowledged.

I. Introduction

The decision to begin or to stop eating is certainly derived from complex processes, involving reactions to external cues like the sensory properties of the available food or the presence of other people who are eating. There is also a host of internal cues derived from the presence or absence of food in the gastrointestinal tract, and from the amount of nutrients circulating in the blood stream or stored in certain tissues. Many factors have actually been identified as contributors to the control of eating: in particular, several peptides and hormones acting peripherally and centrally, and neural feedback loops from the gut which involve activation of different kinds of gastrointestinal chemo- and mechanoreceptors as well as hepatic chemoreceptors. In addition, the hedonic evaluation of the sensory properties of food, depending on the physiological condition and on previous experience, can potently affect the individual's response to the presentation or ingestion of food.

A critical evaluation of all these factors is certainly beyond the scope of this review. We intend to focus on the metabolic control of eating and, in particular, on the role of metabolic factors in the short-term control of eating. The long-term control of eating by body weight or body fat has recently been extensively reviewed [1]. We will first touch on the interactions between eating and energy balance and summarize the evidence for the existence of a postabsorptive metabolic component of eating control. We will then discuss the general characteristics and implications of eating patterns with particular emphasis on human meal-taking behavior. In the main sections we will deal with the possible nature of metabolic cues for hunger and satiety and their registration by sensory neurons in liver and brain. Finally, we will proceed to the integration of metabolic cues and will briefly discuss their modulating effect on taste perception.

II. Eating and Energy Balance

Tissue metabolism requires a continuous supply of utilizable fuels together with oxygen. The fuels that are obtained through periodic eating are continuously released into the circulation by different energy stores. These include the gastrointestinal tract, in particular the stomach, liver glycogen and the fat depots. Under ad libitum eating conditions, fuels absorbed from the gastrointestinal tract are the most important source of energy. If nutrients absorbed from the gastrointestinal tract do not cope with the energy requirements, glucose is released from glycogen and fatty acids and glycerol are released from the fat depots to enter the circulating pool of fuels. The various energy stores clearly differ in capacity and turnover rate. Energy intake may average 8,000–10,000 kJ/day in a normal human being whose fat depots, which represent the major energy store in mammals, are 600,000 kJ or about 75 times the daily energy intake. On the other hand, intracellular ATP provides an energy store with an extremely low capacity, but high turnover rate. The utilization and restoration of energy stores is continuously regulated by autonomic mechanisms [2].

Maintenance of energy balance requires that, on average, energy intake and energy expenditure are equal. Energy equilibrium is of course not achieved from meal to meal. However, in the long-term healthy humans and animals are capable of maintaining body weight (i.e. the best index of energy balance) despite considerable fluctuations in energy intake and energy expenditure. This implies that energy balance is well regulated. Thus, when body weight is forced away from its normal level by various manipulations, animals display compensatory adjustments in voluntary food intake. This sometimes, together with metabolic adaptations, tends to restore body weight to normal [e.g. 3–6]. Furthermore, long-lasting increases in energy needs consistently lead to a compensatory hyperphagia, as is normally the case during lactation [e.g. 7] or during cold exposure [e.g. 8]. Recent studies in humans demonstrate that the increase in heat production which occurs in response to overeating accounts for only about 20% of the amount of energy ingested above energy equilibrium [9–11]. These data indicate that energy balance in humans is regulated mainly by adjustments in voluntary food intake.

Considering the feedback loop between energy balance and eating, it seems clear that nutrient availability or some measure of energy in metabolism affects eating. This assumption is supported by numerous studies which show that parenteral administration of various fuels or pharmacological manipulation of fuel utilization affects eating in man and animals (see section IV). Total parenteral nutrition provides another means to study the importance of metabolic factors involved in hunger and satiety in the absence of oral and gastrointestinal stimulation. In a recent study in humans, 29 generally healthy patients on long-term intravenous alimentation reported only slight hunger or no hunger over most of the observation period of more than 8 months [12]. The patients were in good energy balance with stable body weights. Low levels of appetite that were reported just before occasional oral ingestion of food were attributed to expectancy and hedonic factors rather than energy need because several patients expressed a strong desire for salty and crunchy food [12]. In demonstrating that postabsorptive metabolic cues are sufficient to markedly attenuate hunger sensations even over longer periods of time, these results emphasize the significance of metabolic factors in the control of human eating behavior.

III. Meal-Taking Behavior

Eating in man and animals occurs periodically. Even when food is continuously available, bouts of eating called meals alternate with periods of noneating (intermeal intervals = IMIs). The size, duration, frequency and circadian distribution of meals are important measures of eating behavior, which vary considerably between species. Adult humans normally eat 3–5 meals/day. A similar low meal frequency is only observed in adult pigs, which eat about 3 meals/day [13, 14]. In other species and in young pigs, meal frequency over 24 h is considerably higher. When food is freely available, the dog eats about 6 meals/day [15], whereas rhesus monkey [16], rat [17, 18], cat [19], horse [20], and ruminants [21–23] eat about 8–12 meals/day. The rabbit and the guinea pig may even eat as many as 30 meals/day [24, 25]. In many species a diurnal rhythm is superimposed on this 24-hour meal frequency to the effect that meals are more frequent and in some cases also bigger during the active phase of the diurnal cycle [15–17, 21, 23, 25]. As a result, 2/3 to 3/4 of total daily food intake are normally eaten during the active phase. Interestingly, no rhythm greater than 24 h was found in rats [26], indicating that energy balance is regulated within 24 h through 12- to 12-hour compensations.

Under ad libitum feeding conditions, a significant positive correlation between meal size and duration of the subsequent IMI (postmeal correlation) has often been observed in the rat [17, 18, 27–29] and also in some other species [14–16, 23, 25]. The existence of a postmeal correlation indicates that food intake is regulated from meal to meal, with a meal-related postingestive factor contributing to the maintenance of postprandial satiety.

In the rat, where meal patterns have been investigated most extensively, the postmeal correlation appears shortly after weaning [30], requires a minimum energy intake in a meal of approximately 1.7 kJ [31] and is especially high during the night [29]. In addition, an evolution of the relationship between energy intake at a meal and duration of the subsequent IMI has been observed throughout successive meals during night and day [29]. Simultaneous recordings of eating patterns and O_2 consumption revealed a positive meal-to-meal energy balance during the night, when meals are generally initiated sooner than might be predicted from the ongoing oxidation of the energy-yielding nutrients ingested in the preceding meal. The satiating potency of the ingested energy even decreased with the time elapsed since dark onset [29]. In contrast, a negative meal-to-meal energy balance, which decreased with time, was evident during the day. These data led the authors to conclude that a dual utilization of energy ingested during a nocturnal meal, i.e. current energy metabolism plus fat storage, and a dual source of fuel during the day, i.e. ingested food plus mobilized fat, determines the time of meal onset during night and day, respectively, in freely feeding rats [29]. Reports of a lower satiating potency of ingested or intragastrically infused food during the night compared to the day [32, 33] are in line with this interpretation. Similarly, humans seem to obtain less satiety from a given amount of food in their active phase [34]. In general, the results of LeMagnen and his group suggest a prominent role for metabolic factors in the control of eating and in particular in the maintenance of postprandial satiety. Further support for this assumption is derived from findings indicating that both experimentally induced inhibitions of fuel utilization and parenteral administrations of metabolic fuels often affect meal frequency more than meal size [e.g. 35–39].

Despite some failures to obtain significant correlations between meal size and duration of postmeal intervals [24, 40, 41], the widely held view is that there is a postmeal correlation in some species under certain conditions, and that this can provide some insight into the mechanism of the short-term control of meal size and meal frequency.

Contrary to animals under ad libitum conditions, the duration of postmeal intervals in humans is normally not related to the size of the preceding meal [34, 42–44]. This suggests that when the natural socioecological constraints on human eating behavior are present, meal initiation in humans is fairly independent of the internal state of nutrient depletion. Hence, humans seem to initiate eating in response to an externally determined or learned schedule [44–46]. Under these conditions, however, meal size is positively correlated with the duration of the premeal interval [43, 44, 47], i.e. the internal state at meal onset determines the amount of food eaten. As a consequence, the matching of food intake to energy requirements is then accomplished primarily through variations in meal size. A premeal correlation has also been observed in laboratory animals on a meal-feeding schedule [31] or after food deprivation [4] and in ruminants under some conditions [22, 23]. Interestingly, however, a significant postprandial correlation appears in isolated humans deprived of time cues, when the external stimuli which normally trigger a meal are absent [47–49]. Recent studies demonstrated that the presence of other people does not only eliminate the postmeal correlation [49], but also increases the size and duration of spontaneous meals in humans [50], thus emphasizing the important role of social factors in meal-taking behavior. In one study, both a significant pre- and postmeal correlation were observed [47]. These findings indicate that environmental factors disrupt the original, presumably more physiological, postprandial eating pattern in humans. Furthermore, the results demonstrate that pre- and postmeal correlations are not mutually exclusive features of completely different mechanisms of eating control. Rather they seem to reflect adaptations to differences in the environment.

Further insight into the control of eating behavior of man and animals can be gained if detailed and sensitive measurements of the ingestive process itself, i.e. eating rate or numbers of chews and swallows, can be made. Initial eating rate in a meal apparently reflects both hunger and palatability, while the decline in the rate of ingestion as the meal proceeds reflects the onset of satiety [51, 52]. As a consequence, manipulations that decrease the eating rate normally lead to a decrease in meal size and an increase in meal frequency [e.g. 53]. One explanation for the decrease in meal size with decreased eating rate is that a slower intake rate may allow satiety to develop more efficiently during the course of a meal. An alternative but not mutually exclusive explanation is based on the assumption that eating, once initiated, is sustained by a positive feedback mechanism; hence, a slower intake rate may disrupt this positive feedback that sustains the meal. Meyer and Pudel [54] reported that in contrast to the decrease in eating rate usually observed in normal weight subjects, within meal eating rate does not decrease in obese and latent obese subjects. The authors concluded that this reflects a deficient physiological satiety mechanism in obese people. Others, however, could not confirm these findings in obese subjects [55–57]. A lack of a decrease in the eating rate was confirmed for latent obese people, but was mainly attributed to the fact that cognitive restraint allows these subjects to finish a meal arbitrarily before enough food has been consumed to produce internal feedback and, hence, a decrease in the eating rate [57].

In summary, detailed analysis of meal-taking behavior yields important information concerning the control of eating in man and animals under various environmental conditions and also provides evidence for the existence of a postabsorptive metabolic component of eating control. Further, it demonstrates that part of the control is achieved from meal to meal, whereas energy balance over a longer range is kept constant within about 24 h.

IV. Effect of Circulating Fuels on Eating

We will now focus on the afferent limb in the regulation of food intake, i.e. on the nutrient, hormonal and neural inputs, which provide the controller in the brain with information about nutrient intake and the state of nutrient stores. Particular emphasis in this section will be on the effects of circulating fuels and their utilization on eating. In this context, changes in possible metabolic determinants of hunger and satiety in relation to meal-taking will also be considered.

The circulating energy pool depends on the influx of fuels from the three main energy stores, i.e. gastrointestinal tract, liver glycogen and fat deposits, which clearly represent the main stores of energy. The pattern of fuels absorbed from the gastrointestinal tract depends on the composition of the ingested food and therefore varies, whereas the fuels released from liver glycogen (glucose) and the fat depots (fatty acids and glycerol) do not change in their pattern. In general, glucose, fatty acids and ketones are the main metabolic fuels in most species. However, other nutrients like amino acids, and metabolites such as lactate and pyruvate, also play a role in intermediate metabolism and may affect eating by their availability in the circulation.

A. Glucose

Glucose is used preferentially by most tissues and is almost the exclusive fuel of brain cell metabolism under normal conditions. Considering the prominent role of glucose in metabolism, it is tempting to speculate that some measure of glucose availability controls eating. In the early 1950s, Mayer [58] postulated the existence of chemoreceptors, that have a special affinity for glucose and are activated by the utilization of glucose. According to this glucostatic theory, an increase or decrease in glucose utilization would serve as stimulus for satiety or hunger, respectively [58].

This hypothesis was based in part on the observation that carbohydrate reserves are proportionally much more depleted between meals than are reserves of protein or fat [58].

In concordance with the glucostatic theory, it has often been shown that cellular glucoprivation stimulates eating. Stimuli that provoke this state include glucose animetabolites like 2-deoxy-D-glucose (2-DG) or 5-thioglucose, which act as competitive inhibitors of the phosphohexoseisomerase [59, 60], and high doses of insulin, which initially increase but later decrease glucose utilization because of their hypoglycemic effect. The effects of insulin on eating will be discussed in a later section.

Parenterally administered 2-DG stimulates eating in many species [61–64] including man [65, 66]. In addition, humans reported increased hunger ratings in response to 2-DG administration [65]. The eating response to 2-DG proved to be greater in satiated than in hungry rats [67], which is presumably related to a higher rate of glucose utilization in satiated compared to hungry rats. Furthermore, 2-DG increases food intake by primarily increasing meal frequency [36, 37], thus implicating glucose utilization in the maintenance of postprandial satiety. Inhibition of glucose utilization seems to selectively increase carbohydrate ingestion because 2-DG stimulated carbohydrate intake but decreased fat and protein intake in rats with access to separate sources of the three macronutrients [68]. In addition to its behavioral effect, 2-DG provokes compensatory sympathoadrenal responses, like stimulation of glycogenolysis and inhibition of insulin secretion [69, 70], which result in hyperglycemia [59, 61]. These responses help to conserve glucose for the function of the insulin-independent brain tissue.

Although it is generally assumed that 2-DG-induced eating reflects cellular glucoprivation, the exact relationship between glucoprivation, blood glucose level and eating is not yet completely understood. Ritter et al. [71] found that 2-DG-induced eating still occurs when food is withheld 6 or 8 h after 2-DG administration, until the glucoprivic stimulus has presumably dissipated. The delayed eating response to 2-DG was diminished or prevented by various metabolic fuels that were infused or ingested prior to the presentation of food [72, 73]. These findings suggest that the stimulation of eating by 2-DG is not exclusively and directly coupled to acute glucoprivation.

There is evidence that a decrease in blood glucose levels is among the physiological signals that trigger a meal. With continuous recording of blood glucose, Louis-Sylvestre and LeMagnen [74] observed a small (6–

11 %) drop in blood glucose just before spontaneous meals in rats. A similar premeal decline in blood glucose was previously observed in rats adapted to a schedule of two daily meals [75]. More recent studies showed that intravenous glucose infusions, which attenuated the premeal decline in blood glucose, delayed the onset of the meal [76]. Furthermore, when food was withheld until blood glucose levels had returned to normal, meal onset was also delayed [76]. These results suggest that the premeal decline in blood glucose in the rat is causally related to meal onset. Unfortunately, it is unclear whether a change in glucose utilization accompanies the premeal decline in blood glucose because recordings of the respiratory quotient in relation to spontaneous meals yielded conflicting results [37, 77].

A recent attempt to detect a similar robust premeal decline in blood glucose in humans failed [78]. This may, however, be due to comparatively long intersample intervals of 20 min in the respective study [78]. In another study, in which blood was sampled in 4-min intervals, several subjects showed a drop in blood glucose level just before a scheduled lunch [79]. Frequent postprandial blood sampling (at 4-min intervals) in the same study revealed pronounced postprandial oscillations in plasma glucose and insulin levels [79]. Both long-term (mean period 51–112 min) and short-term (mean period 9–14 and 20–30 min) oscillations were observed. All oscillations were highest after meals and were then damped, reverting to basal glucose levels at about 340 min after the meal [79]. These oscillations are presumably related to cyclical variations in pancreatic secretion and to the biphasic pattern of the appearance of ingested glucose in the blood stream which has been observed in man [80]. Moreover, the second or third nadir in postprandial blood glucose seems to coincide with the cessation of glucose absorption from the gut [80] and hence with the transition from exogenous glucose delivery to endogenous glucose production, which occurs at about 3 h after the meal [81]. Thus, considering the common duration of intermeal intervals in humans, a causal relationship between decreases in blood glucose levels and hunger seems possible. Indirect evidence for a causal relationship between decreases in blood glucose levels or glucose availability and hunger in man is also derived from studies in which meals with slow-release carbohydrates were shown to produce a smaller but clearly longer sustained rise in blood glucose and a decrease in hunger ratings compared to meals containing potatoes [82]. Finally, a respiratory quotient of about 0.82 and, hence, low carbohydrate oxidation was also observed just prior to breakfast in healthy men [83].

Blood glucose levels in monogastric species including man increase during and after carbohydrate-containing meals [e.g. 74, 79, 84]. However, in contrast to the rather constant increase in food intake in response to an inhibition of glucose utilization, attempts to induce satiety by parenteral applications of glucose have not always been successful. While Mayer [58] and many other authors [e.g. 85–91] observed a marked suppression of eating after parenteral administration of glucose, some did not [e.g. 92, 93]. Closer examination of the respective data suggests that this discrepancy is due to different experimental conditions, like the timing or route of glucose application, the composition of the test diet or the duration of the feeding test. Glucose has also been shown to reduce food intake after intragastric administration when food was offered 2–3 h later [94]. This delay was sufficient for absorption of the glucose load to occur. More recently, intravenous infusions of glucose plus amino acids in rats, given only during the period when the rats were fed, reduced food intake by about 70% of the infused calories, whereas intravenous infusion of fat reduced food intake by only about 17% of the infused calories [95]. In contrast, continuous infusions of glucose over 24 h often had a smaller effect [35, 93, 96] or no effect on food intake at all [92]. This suggests that intravenous glucose infusions are more satiating at a time when concomitant oral food intake stimulates other satiety signals which interact with and strengthen the glucose effect. A potentiation of parenterally administered glucose's satiety effect by concomitant carbohydrate intake has been observed by Novin et al. [90]. Interestingly, in some studies the satiating potency of glucose increased significantly when insulin was infused simultaneously [35, 97]. This provides indirect evidence for the role of glucose utilization in the satiety effect of parenterally administered glucose and suggests that the involved glucoreceptors are at least partly sensitive to insulin. Finally, a causal relationship between increases in glucose utilization and the development of satiety during meals is possible because glucose utilization increases during meals in rats [77] and man [83, 98].

In summary, there is considerable evidence for a postabsorptive metabolic effect of glucose on eating in addition to any effect that ingested glucose may have on the gastrointestinal tract.

B. Nonesterified Fatty Acids

It has been proposed that circulating fuels whose plasma levels reflect the level of adiposity contribute to the long-term control of food intake [7]. Fat rather than carbohydrate or protein is used or stored the most in response to day-to-day fluctuations in energy intake and energy expenditure [99]. Changes in plasma concentration and utilization of nonesterified fatty acids and perhaps glycerol have been implicated in the circadian variations in food intake [26, 100]. According to this hypothesis, high plasma levels of fat metabolites (especially fatty acids) resulting from diurnal lipolysis would inhibit eating. A powerful hypophagic effect of intravenously infused lipids was actually observed in one study [89]. In general, however, intravenous infusions of fatty acids or lipids yielded mixed results [35, 95, 101–103] and the occasionally observed effects of these infusions on eating were rather small. High plasma levels of nonesterified fatty acids have even been considered as hunger signal [104, 105]. However, because fat mobilization during fasting represents an attempt of the organism to compensate for the lack of exogenous energy, it appears reasonable to assume that a fasting individual without increased plasma levels of fat metabolites would be even more hungry. Interestingly, hunger appears to decrease as soon as mobilization of depot fat is fully activated during food deprivation [106]. Thus, it is more plausible that a high rate of fatty acid oxidation inhibits eating. Oxidation of fatty acids seems to be sensed by metabolic sensors involved in the control of eating and seems to be particularly important for food intake control when a fat-rich diet is ingested. We found that inhibition of β-oxidation of fatty acids by mercaptoacetate is associated with increased food intake in rats fed a fat-rich diet (18% fat, w/w), but not in rats fed a low fat diet (3.3% fat, 77% starch) (fig. 1) [107]. Furthermore, the hyperphagic effect of mercaptoacetate was more pronounced during the day than during the night, probably because lipolysis and fatty acid oxidation in the rat is accelerated during the day [108]. Interestingly, mercaptoacetate stimulated carbohydrate as well as fat intake in rats selecting from equicaloric high fat (40% fat) and low fat (3.3% fat, 77% starch) diets, indicating that inhibition of fatty acid oxidation generates an unspecific hunger signal (fig. 2). Mercaptoacetate blocks fatty acid oxidation by inhibiting the acyl-CoA-dehydrogenases located in the mitochondrial matrix and therefore impairs mitochondrial β-oxidation of fatty acids [109, 110]. Further evidence for the importance of mitochondrial oxidation in the satiety effect of fatty acids is derived from findings showing that methylpalmoxyrate, an inhibitor of carnitine palmitoyltransferase, increased food intake in rats fed a diet high in long-chain fatty acids, but did not affect food intake in rats fed a diet high in medium-chain fatty acids, which do not require the enzyme carnitine palmitoyltransferase for mitochondrial uptake and oxidation [111]. In order to determine

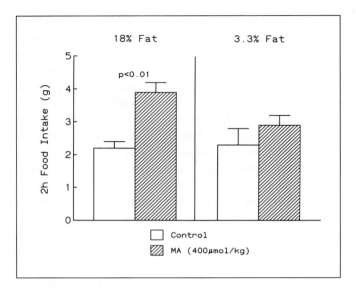

Fig. 1. Differential effect of mercaptoacetate (MA) on food intake in rats fed a fat-rich (18% fat, w/w) or a low fat (77% starch, 3.3% fat) diet. MA was injected intraperitoneally at the onset of the dark phase of the lighting cycle. Each bar represents the mean ± SEM of 27–28 rats. From Scharrer and Langhans [107].

whether the hyperphagia elicited by an inhibition of fatty acid oxidation (lipoprivic feeding) is due to a delay of meal-ending satiety or to a disturbance of postprandial satiety, we recorded the meal pattern of mercaptoacetate-treated rats [38]. Mercaptoacetate increased food intake by reducing the latency to eat after injection and the duration of the subsequent intermeal interval without significantly affecting the size of the first meal. This indicates that fatty acid oxidation contributes to the maintenance of postprandial satiety. It should be mentioned, however, that it is as yet unproven that physiological fluctuations in fatty acid oxidation play a role in the meal-to-meal control of eating.

The potent hyperphagic effect of mercaptoacetate in rats kept on a fat-rich diet is interesting because dietary fat accounts for almost 40% of total daily caloric intake of people in industrialized countries. Consumption of diets high in fat has been associated with increased caloric intake, body weight, and fat deposition in laboratory animals [see 112] as well as

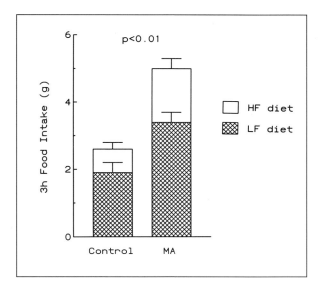

Fig. 2. Effect of mercaptoacetate (MA, 400 µmol/kg) on food intake in rats selecting from an equicaloric high fat (HF, 40% fat) and a low fat (LF, 3.3% fat, 77% starch) diet. MA was injected intraperitoneally at the onset of the dark phase of the lighting cycle. Values are means ± SEM of 28 rats, each. p < 0.01 indicates that MA increased LF and HF diet intake significantly.

in humans [83, 113, 114]. The high fat hyperphagia may be related to the high caloric density of most high fat diets, because often hyperphagia and increases in body weight do not occur when the energy density of the high fat diet is reduced [e.g. 87]. Flatt et al. [83] found that the rates of overall fat or carbohydrate oxidation as evidenced by changes in the respiratory quotient in response to meals in humans were not affected by the fat content of the meal. Therefore, after a fat-rich meal, long-chain triglycerides seem to be channeled towards storage rather than oxidation. This assumption is compatible with findings of a meal-induced increase in adipose tissue lipoprotein lipase activity [e.g. 115], which promotes triglyceride storage. The possibility that the oxidation of long-chain fatty acids is comparatively inefficient as long as other fuels are available, might also explain the rather weak effect of parenterally administered fatty acids or triglycerides on eating (see above). However, a failure of fat ingestion to decrease the respiratory quotient does of course not exclude the possibility that fatty acid oxidation in a particular organ like the liver increases in response to a fatty meal (see also section V). It must also be considered in this context that long-term changes in the nutrient composition of an ingested diet lead to an adaptation of enzymes involved in the degradation of the respective nutrients. Consumption of a high fat diet for instance leads to an adaptation of fatty acid oxidation, as indicated by the increased rate of oxidation of acyl-CoA esters by isolated liver mitochondria [116] and a decrease in glucose oxidation [87]. Accordingly, in subjects adapted to various levels of dietary fat, overall substrate utilization as evidenced by the respiratory quotient closely parallels diet composition [117]. In another recent study, subjects compensated perfectly for the caloric content of a meal regardless of its fat content [118]. These findings are also compatible with a postabsorptive contribution of dietary fat to the metabolic control of eating, because compensation did not occur before 5 h after the meal, at which time ingested fat was certainly oxidized [119]. Thus, most of the available evidence supports the view that fatty acid oxidation plays a role in the metabolic control of eating, at least when considerable amounts of fat are ingested, as is often the case in humans.

Recent findings suggest that glucose and fat metabolism exert a coordinated control over eating behavior. Combined treatment of rats with 2-DG, which inhibits glucose utilization, and methylpalmoxyrate, which inhibits fatty acid oxidation, increased food intake of rats in a synergistic fashion [120]. This synergy was observed at doses of the two agents that alone did not increase food intake [120], indicating that subthreshold changes in glucose and fatty acid oxidation add to produce an eating response. Similar results were obtained with combined administration of 2-DG and nicotinic acid, which blocks fat mobilization [121]. In addition, parenterally and intragastrically administered glucose proved to be more satiating when fatty acid oxidation rate was high [87]. These findings lend strong support to theories suggesting that some general measure of energy flow in metabolism rather than changes in the utilization of a particular fuel controls eating [35, 122–124].

C. Other Fuels

In addition to glucose and nonesterified fatty acids, various other metabolic fuels have been proposed to signal hunger and satiety by their availability in the blood. We found that glycerol, malate, D-3-hydroxybutyrate (DHB), lactate or pyruvate reduced food intake by similar amounts after subcutaneous injection in rats [124–127]. In contrast, dihydroxyace-

tone, oxaloacetate and acetoacetate, which are the immediate oxidation products of glycerol, malate and DHB, respectively, failed to affect food intake significantly [124, 126]. Moreover, the hypophagic effect of glycerol disappeared with high levels of dietary protein [128], which are known to suppress the mitochondrial glycerol dehydrogenase [129]. The effects of lactate and pyruvate disappeared with high levels of dietary fat [124], which inhibit mitochondrial pyruvate oxidation [130]. Collectively, these data suggest that the mitochondrial oxidation of metabolic fuels is crucial for the hypophagic effect. The transient food intake suppression induced by the tested metabolites was always more than caloric compensation for the injected energy. Thus, the findings support theories implicating the rate of energy production in the control of food intake [35, 122, 123] and actually identify particular metabolic steps that appear to be crucial.

As glycerol is released from adipose tissue together with fatty acids during lipolysis, its hypophagic effect after parenteral administration [125, 131–135] may well reflect a physiological role in the control of food intake and energy balance. This assumption has been questioned because glycerol doses which reduce food intake reliably raise plasma glycerol levels transiently beyond physiological levels [135]. It must be considered, however, that lower doses of glycerol might be effective if glycerol was injected together with other metabolites that affect food intake. We have for instance shown that glucose and malate act synergistically regarding their inhibitory effect on eating [39]. Doses of glucose (i.p.) and malate (s.c.) that were ineffective alone dramatically reduced food intake of rats when combined (fig. 3). Higher doses of malate and glycerol presumably reduce food intake through a similar mechanism [124]. As previously mentioned, metabolic inhibitors affecting glucose or fatty acid metabolism also act synergistically with respect to their stimulatory effect on eating [120, 121]. Different hormones have also been shown to inhibit food intake synergistically [136]. Thus, a synergism of various stimuli controlling food intake seems to be an important principle in the control of eating that could explain the frequent finding that unphysiological high doses of putative satiety agents are necessary to affect eating behavior.

An increased rate of fatty acid oxidation is normally associated with an increased production of ketone bodies. Similar to glycerol, DHB reduced food intake in rats after parenteral administration [126, 127]. Furthermore, the hyperphagic response to 2-DG [72] or insulin [137] was also diminished by infusion of DHB. Section of the hepatic branch of the vagus eliminated the hypophagic effect of subcutaneously administered

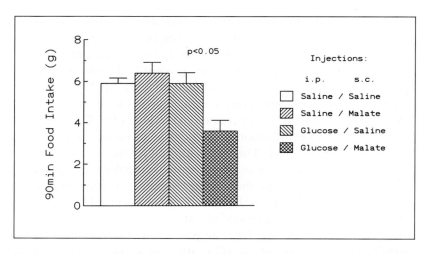

Fig. 3. Synergistic effect of glucose (9 mmol/kg) and malate (7.7 mmol/kg) on food intake. Rats were kept on a 5 h feeding/19 h deprivation schedule. Combinations of intra-peritoneal (i.p.) injections of glucose or saline and subcutaneous (s.c.) injections of saline or malate were given as indicated and 20 min apart. Food was presented immediately after the second injection. Values are means ± SEM of 14 rats, each. From Sydler [39].

DHB [138], suggesting that the liver is involved in the hypophagic effect of exogenous DHB. As acetoacetate in contrast to DHB did not reduce food intake significantly [126], hepatic oxidation of DHB to acetoacetate might be involved in the effect of exogenous DHB on food intake [126, 127]. The hypophagic effect of DHB is presumably not physiologically relevant because normally DHB is formed from acetoacetate in the liver. However, DHB might be oxidized by hepatic vagal afferents and affect food intake in this way. It should also be mentioned that acetoacetate reduced food intake similar to DHB in experiments of others [139]. It is possible that periph-erally administered DHB inhibits eating in part through a central mecha-nism, because it also reduced food intake after intracerebroventricular administration [140]. Again the effect of DHB on food intake may well depend on the plasma levels and utilization of other metabolic fuels.

In contrast to the hypophagic effect of DHB, the hypophagic effects of lactate and pyruvate [124, 131, 141] may be of physiological relevance because plasma levels of lactate have often been shown to increase during carbohydrate meals in various species [142–145], including man [146].

The potential significance of lactate and pyruvate for the short-term metabolic control of eating will be discussed in detail later. Interestingly, recent findings indicate that adipose tissue is an important source of lactate during eating [147, 148]. As adipocyte lactate production varies with the level of adiposity [149], lactate may even contribute to the long-term control of food intake by body fat. Malate, which also reduced food intake after subcutaneous injection [124], is found in the plasma only in low concentrations. Extracellular malate is therefore presumably not involved in food intake control under normal conditions.

As already mentioned, circulating amino acids have also been implicated in the metabolic control of eating because parenteral administration of balanced amino acid solutions or protein hydrolysates has often been shown to reduce food intake in several species including man [92, 150–152]. While ammonia toxicity may contribute to the suppression of food intake induced by high doses of amino acids [152], metabolic effects other than ammonia production seem to be involved at low amino acid doses [152]. The exact mechanism of the hypophagic effect of amino acids is largely unknown, but some evidence suggests that the liver might be involved (see next section). Finally, some authors suggested that changes in the ratio of plasma tryptophan concentration to that of large neutral amino acids controls carbohydrate/protein selection through changes in brain serotonin synthesis [153, 154]. However, this hypothesis is controversial [155, 156].

In summary, the data presented in this section generally support the hypothesis that fluctuations in the utilization rate of different metabolic fuels control eating in a way that the rate of fuel utilization is inversely related to eating [35, 122, 123, 157].

V. Control of Eating by Hepatic Metabolism

A. Evidence for a Role of Hepatic Metabolic Sensors

To affect eating, the utilization of metabolic fuels must be monitored by sensors, which are connected to the CNS circuitry controlling eating behavior. The liver has received much attention as a probable location for those sensors. There is convincing evidence for the existence of various sensory functions of the liver [158, 159]. Hepatic sensory functions may help to maintain homeostasis in mammals and may well contribute to food intake regulation. In fact, it seems plausible that the liver is involved in the metabolic control of eating because of its key role in metabolism and because it is exposed to the nutrient flow from absorption. The liver is able to oxidize almost all metabolic fuels, including glucose, fatty acids, glycerol, and amino acids. Further, the liver stores glucose in the form of glycogen and thus functions as a short-term energy store. Russek, [85] was the first to suggest that the liver contributes to satiety, and subsequently, attention was focussed on the possible role of hepatic glucosensitive metabolic sensors in the control of food intake. Niijima [160] provided electrophysiological evidence for the existence of hepatic glucosensitive vagal neurons. Using the isolated perfused guinea pig liver, Niijima [160] showed that the discharge rate of hepatic vagal afferents decreased when D-glucose was added to the perfusion medium. Later liver perfusion studies, which were also performed in situ, extended these findings in demonstrating that only D-glucose decreased the firing rate of hepatic vagal afferents, while other sugars (D-mannose, D-fructose, D-galactose, L-glucose, D-xylose, D-arabinose) did not [161]. Further, intraportal infusion of 2-DG increased the firing rate of hepatic vagal afferents [162], suggesting that glucose utilization is sensed by the hepatic glucosensors.

Several lines of evidence imply that the hepatic glucosensors are involved in the control of eating. Russek [163] reported that intraportal but not intrajugular infusion of glucose suppressed food intake in 22h fasted dogs. Attempts to replicate these original findings and to extend them to other species yielded equivocal results. Thus, portal glucose infusion has been shown to decrease food intake [e.g. 86, 88, 90, 164], to have no effect [e.g. 84, 132, 165] or even to increase food intake under some conditions [166]. However, more recent systematic tests of portal glucose infusions by Tordoff and Friedman [91, 167, 168] convincingly demonstrated that glucose infusions that produced metabolic changes within the physiological range and did not affect systemic blood glucose levels reduced food intake. Moreover, when glucose or control infusions were paired with the presentation of flavored food, rats given intraportal glucose infusions that reduced food intake developed a preference for the flavor that had been paired with glucose infusion [91]. This demonstrates that intraportally infused glucose is not aversive but rather forms the basis for the acquisition of a learned food preference. In addition, the data suggest that the failure of intraportally infused glucose to reduce food intake in some studies was due to the test conditions, which did not control for a powerful carryover effect of previous portal glucose infusions [91]. Interestingly, food intake was reduced by the same amount irrespective of the concentration of the glucose solution and the amount of glucose infused intraportally [168]. Therefore, the critical metabolic event in the liver that influences food intake seems to be only indirectly coupled to portal glucose levels [168]. Other aspects of the glucose supply like the portal-arterial glucose gradient [168] or the rate of glucose delivery into the portal vein [169] may be crucial. Thus, while food intake seems to be reduced when glucose is delivered at rates that match the physiological norm [168], meal size, meal duration and cumulative food intake seem to be increased when glucose or other hexoses are delivered at a higher rate [169]. In an attempt to explain this paradoxical stimulation of eating, Novin et al. [169] performed fast and slow intraportal infusions of double-labelled (^{14}C and ^{3}H) glucose and fructose in rabbits and found that at the slow infusion rate (=inhibition of eating) ^{14}C uptake in liver mitochondria was twice as much as at the high infusion rate (=stimulation of eating). In contrast, the uptake of the ^{3}H label into lipid for the fast infusion was twice that of the slow infusion [169]. Although the mechanisms of this diversion remain to be clarified, the results indicate that hepatic mitochondrial oxidation is involved in the satiety effect of infused glucose and fructose. A practical implication of these findings for the control of human eating behavior is also conceivable. The high infusion rate might be analogous to fast consumption of soft drinks and candy bars which are rapidly digested and absorbed, whereas the slow infusion rate might mimic the consumption of complex carbohydrates, delivering glucose more slowly [169].

Some evidence for a hepatic glucosensitive mechanism in the control of food intake is also derived from studies with 2-DG. Infusion of 2-DG into the hepatic portal vein in rabbits caused a more rapid and greater increase in food intake than infusion into the jugular vein and the effect was reduced by vagotomy [63]. In similar experiments in rats, both intrajugular and intraportal infusions of 2-DG produced a similar increase in food intake [170]. Moreover, hepatic branch vagotomy did not reduce the stimulatory effect of 2-DG on eating in rats during the day [171]. Delprete and Scharrer [172] recently confirmed this finding, but additionally observed that the eating response to 2-DG was reduced by hepatic branch vagotomy, when 2-DG was injected in the early dark phase of the lighting cycle and especially after consumption of a test meal. The finding that the disruption of postprandial satiety by 2-DG is partially dependent on an intact hepatic branch of the vagus supports the hypothesis that glucose utilization by hepatic vagal afferents or hepatocytes contributes to the maintenance of postprandial satiety. Finally, it is also interesting in this context that complete subdiaphragmatic as well as hepatic branch vagotomy disrupted the otherwise reliable coupling between the premeal decline in blood glucose levels (see section III) and meal initiation in the rat. Thus, while the transient premeal declines in blood glucose in both vagotomized groups were quantitatively similar to those observed in intact rats, they predicted meal initiation in only 55% of all trials compared to the 100% coupling of these events in intact rats [173]. These findings are consistent with the idea that hepatic glucose sensors play a role in the detection of the transient premeal decline in blood glucose and the reliable mapping of this decline into meal initiation [173].

Hepatic vagal afferents project to the nucleus of the solitari tract (NST) [174], which represents an important relais for visceral and gustatory afferents involved in the control of food intake [175, 176]. From the NST the information can proceed rostrally to the hypothalamus directly or indirectly through the parabrachial nucleus. NST neurons and lateral hypothalamic neurons also respond to glucose in the same way as hepatic vagal afferents [177, see next section]. Furthermore, the lateral hypothalamic glucosensitive neurons also respond to intraportal glucose application [178], suggesting that glucosensitive hepatic vagal afferents are linked to the lateral hypothalamus. In a previous study, however, the effect of intraportal glucose on the firing rate of lateral hypothalamic neurons could be blocked by splanchnic nerve section [179]. Therefore, the hepatic glucosensors might also be linked to the lateral hypothalamus via splanchnic afferents. Like vagal afferents, the majority of hepatic splanchnic afferents projects to the NST [180]. Collectively, the electrophysiological data at hand indicate that the glucosensitive neurons in liver, NST and lateral hypothalamus are anatomically and functionally related and seem to represent a network that senses the availability of glucose and may be involved in glucose homeostasis as well as food intake regulation.

In addition to glucose utilization, fatty acid oxidation may also affect eating in part through hepatic sensors because the hyperphagia of rats fed an 18% fat diet in response to mercaptoacetate was partially blocked by hepatic branch vagotomy [181]. In a total of five experiments, the stimulatory effect of mercaptoacetate on eating was greater in sham-vagotomized rats than in vagotomized rats. There was also a small effect of mercaptoacetate on eating in vagotomized rats, which was statistically significant in one of two experiments performed in the middle of the bright phase of the lighting cycle. At this time of the diurnal cycle the eating response to mercaptoacetate was maximal in the rats with an intact hepatic vagus

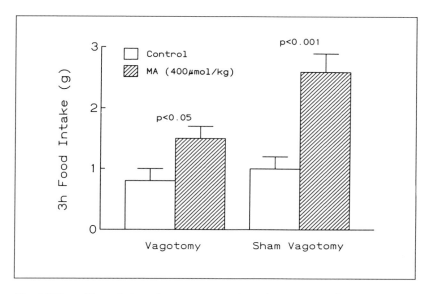

Fig. 4. Effect of hepatic branch vagotomy on the eating response of rats to mercaptoacetate (MA). Rats were fed an 18% fat diet and MA was injected intraperitoneally in the middle of the bright phase of the lighting cycle. From Langhans and Scharrer [181].

branch. The results of this experiment are shown in figure 4. The residual eating response to mercaptoacetate in rats with hepatic branch vagotomy suggests that extrahepatic sensors may in part be involved in mercaptoacetate's effect on food intake, or that hepatic afferents that are spared in hepatic branch vagotomy contribute. As already mentioned, hepatic metabolic sensors may also be linked to the brain through splanchnic afferents [179]. Since mercaptoacetate retained its potency to stimulate eating after peripheral blockade of cholinergic transmission by atropine methylnitrate [181], fatty acid oxidation seems to affect eating at least in part through vagal afferents and not efferents. More recently, Ritter and Taylor [182] investigated the effect of mercaptoacetate on food intake in capsaicin-treated rats fed a fat-rich diet. Capsaicin is a toxin that destroys fine-diameter unmyelinated primary sensory neurons, including many visceral afferents. Capsaicin-treated rats did not eat in response to mercaptoacetate, whereas vehicle-treated rats clearly did [182]. Moreover, eating could not be elicited by injection of mercaptoacetate into the lateral or fourth ventricle of rats. Finally, lipoprivic eating was abolished by subdiaphragmatic vagotomy as well as by lesions of the vagal sensory terminal fields in the area-postrema NST [183]. All these findings are consistent with the assumption that the metabolic sensors involved in lipoprivic eating reside in the liver and maybe other viscera. It is also noteworthy in this context that the liver obtains the largest fraction of its ATP through oxidation of fatty acids [184].

The hypophagic effects of glycerol, lactate, pyruvate, malate and DHB (see previous section) also seem to be mediated at least in part by hepatic sensors which are linked to the brain through the vagus, since subcutaneous injection of these metabolites failed to suppress eating in rats with hepatic branch vagotomy [138].

In summary, under certain conditions, hepatic branch vagotomy attenuates at least some of the changes in food intake that can normally be observed in response to various metabolic manipulations. These results are compatible with the assumption that metabolic cues, which are sensed in the liver and conveyed to the brain through the vagus, can affect eating. Nevertheless, data obtained after hepatic branch vagotomy should be interpreted cautiously because the hepatic branch of the vagus does not innervate only the liver and is not the only afferent neural connection between liver and brain [176].

B. Changes in Hepatic Metabolism Related to Eating

In order to understand the proposed role of hepatic metabolism in the control of eating, it is useful to consider the profound changes that can be observed in the supply of metabolites to the liver as well as in hepatic metabolism before, during and in response to eating. The average adult person in industrialized countries ingests a diet that provides about 13–15% of the energy from proteins, 40% from carbohydrates, and almost 40% from fat. Approximately 7% of the energy is derived from alcohol. A considerable part of the carbohydrates is ingested in the form of sucrose. Thus, after digestion and absorption of foodstuffs, the liver is presented through the portal circulation with a mixture of amino acids, monosaccharides, medium-chain fatty acids, and sometimes, ethanol.

Blood glucose levels and glucose oxidation have often been shown to increase in response to carbohydrate-containing meals (see previous section). However, the relative importance of the liver compared to peripheral tissues in the disposal of ingested glucose is not so clear. Although in the postabsorptive state the fractional extraction of portal vein glucose by the liver is normally rather low in man [80, 185], the total amount of ingested glucose that ends up in the liver seems to be considerable [146, 185]. These seemingly contradictory observations are explained by the fact that in vivo most glucose apparently enters hepatic glycogen by an indirect route which requires transformation of ingested glucose into three-carbon units by extrahepatic tissues [186–189]. One reason for the low first-pass hepatic uptake of ingested glucose seems to be the low phosphorylating capacity of hepatocytes for glucose [190, 191]. The activity of hepatic glucokinase, the specific enzyme responsible for channeling the portal flux of absorbed glucose into hepatic glucose uptake and glycogen storage, seems to be particularly low in man [192]. According to recent studies, a greater proportion of hepatic glycogen may be formed directly from the uptake and phosphorylation of portal glucose [193] in the rat, where hepatic glucokinase activity is higher than in man [192].

In the ad libitum fed rat, spontaneous meals during the dark phase of the lighting cycle do not increase portal vein blood glucose level significantly [194]. This may be due to a continuously high rate of glucose absorption from the intestine in the ad libitum condition, when rats eat about 10–12 meals/day and mostly during darkness (see section III). In a recent study in man, frequent sipping of a liquid formula diet over a 12-hour period also eliminated the meal-related fluctuations in blood glucose levels that were seen when the same amount of the formula diet was given in three meals [195]. On the other hand, even mild food deprivation is sufficient to observe a meal-induced increase in portal vein blood glucose in the rat [84, 194, 196]. Consequently, meals in mildly food-deprived rats may be well suited for investigations in the physiological changes accompanying normal eating in man. In rats food deprived for 12 h during the bright phase, where food intake is normally low, portal vein glucose levels increase profoundly during the first meal after refeeding [194, 196]. In addition, eating stimulates a transient hepatic glycogenolysis in 12-hour food-deprived and in ad libitum-fed rats [194, 196]. This prandial hepatic glycogenolysis and concomitant glucose release together with the increase in portal plasma glucose may combine to provide more glucose for the hepatic glucose sensors possibly involved in the production and maintenance of satiety.

In addition to the prandial hepatic glycogenolysis, considerable hepatic glycolysis seems to occur in response to eating after mild food deprivation in dogs [197] and rats [198], as well as in response to oral glucose ingestion in humans [146]. Findings of in vitro experiments complement this notion in demonstrating that liver cells from fed mice respond to

glucose in the incubation medium with increased glycolysis, whereas glycogen synthesis prevails in response to glucose in liver cells from mice
fasted for at least 24 h [199]. In line with the latter findings, the hepatic
concentration of fructose-2,6-biphosphate remains low for several hours
after refeeding in previously fasted animals, and rises only when glycogen
has been largely repleted [190]. Low levels of fructose-2,6-biphosphate
favor gluconeogenesis from gluconeogenetic precursors, whereas a rise in
fructose-2,6-biphosphate levels triggers glycolysis [190, 200]. These findings might explain why hepatic glycolysis, with or without concurrent
hepatic gluconeogenesis, apparently occurs during and immediately after
eating under normal conditions in man and animals [146, 191, 198, 201].
According to the concept of 'metabolic zonation' of liver metabolism [202,
203], glycolysis, and concurrent gluconeogenesis and glycogen synthesis,
may occur in perivenous and periportal hepatocytes, respectively. Considering the evidence for a role of glucose utilization in the metabolic control
of eating (see section IV), a physiological increase in hepatocellular glucose
degradation in response to eating may contribute to the production and
maintenance of satiety in addition to the putative effect of glucose utilization in hepatic vagal afferents (see previous part of this section).

Lactate and pyruvate may also be involved in the short-term control of
food intake. Although most of the ingested glucose seems to be absorbed
intact [204], an increase in portal vein plasma lactate concentration in
response to intragastric glucose loads has been observed in unrestrained,
chronically catheterized rats [142–144] and dogs [205]. As already mentioned (see section IV), some of the lactate is probably derived from the
adipose tissue [148]. Eating after mild food deprivation in rats is also
accompanied by an increase in portal vein and hepatic lactate concentration [145]. As shown in figure 5, the portal vein lactate concentration rises
faster than hepatic vein or aortal lactate concentration. In addition to the
increased portal supply of lactate, degradation of glucose derived from
hepatic glycogenolysis may contribute to the prandial increase in hepatic
lactate content (fig. 5) as suggested by Davis et al. [197]. At physiological
plasma lactate concentrations, lactate uptake by the liver seems to be
mainly carrier-mediated and is probably not rate-limiting for hepatocellular lactate utilization [206]. Lactate is a dead end in metabolism, but is
easily converted to pyruvate for further utilization [190]. As already mentioned, hepatic mitochondrial oxidation of pyruvate seems to be involved
in lactate's hypophagic effect [124]. An increased supply of lactate to the
liver in response to eating may lead to an increased hepatic oxidation of

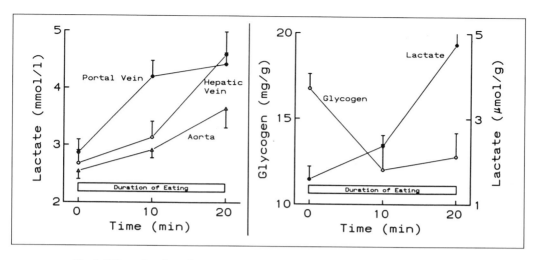

Fig. 5. Effect of eating after mild food deprivation on portal vein, hepatic vein and aortal plasma lactate levels (left) and on hepatic concentrations of glycogen and lactate (right). Each point represents the mean ± SEM of 9 rats. From Langhans [145].

lactate and pyruvate and may therefore contribute to the production and maintenance of satiety. Lactate or pyruvate may also interact with glucose to inhibit eating (see section IV).

Changes in portal vein and hepatic vein lactate and pyruvate levels of course do not necessarily reflect the hepatic metabolism of these metabolites. Nevertheless, there is evidence that eating leads to an increase in hepatic oxidation of pyruvate that might contribute to satiety. Thus, pyruvate dehydrogenase activity immediately increases with eating in mildly food-deprived rats [207]. Interestingly, this increase in response to eating is markedly delayed in 24- or 48-hour fasted rats [207]. The flux of pyruvate through the mitochondrial pyruvate dehydrogenase or pyruvate carboxylase pathways depends on the rate of mitochondrial pyruvate uptake [208, 209] and may be stimulated by fatty acid oxidation under certain conditions. Fatty acid oxidation is generally believed to inhibit mitochondrial pyruvate dehydrogenase [see 210]. However, this holds true especially for high pyruvate levels [see 210], whereas fatty acid oxidation apparently stimulates pyruvate dehydrogenase when pyruvate levels are within the physiological range [209, 210]. Interestingly, considerable hepatic uptake of nonesterified fatty acids and DHB output still occur during eating after mild food deprivation in the rat, at the time when hepatic lactate uptake is maximal [145]. Thus, it is tempting to speculate that under these conditions the continuing, although diminishing oxidation of fatty acids during eating might promote hepatic mitochondrial pyruvate utilization in response to the profound increase in lactate supply to the liver. This might contribute to the gradual development of satiety that is reflected by the slowdown of eating observed before eating actually stops [51, 52]. While most long-chain fatty acids get to the liver after lipolysis of triacylglycerols in the adipose tissue, medium- and short-chain fatty acids derived from dietary fat are easily transferred to the liver through the portal vein [184]. Thus, upon digestion and absorption of a meal containing fats, the liver is presented with at least some medium-chain fatty acids which feed into the mitochondrial fatty acid oxidation pathway. A rapid and tremendous increase in blood ketone body levels has actually been observed after ingestion of medium-chain triglycerides in man [211]. Therefore, these findings are also compatible with a role of hepatic fatty acid oxidation in meal-ending satiety. There is evidence that metabolic factors are primarily involved in the control of meal frequency instead of meal size [e.g. 38, 212]. However, this does not necessarily exclude the possibility that metabolic cues, perhaps in concert with gastrointestinal cues, also contribute to meal-ending satiety, at least under certain conditions. For instance, the hypophagia observed after complete absorption of intragastrically infused glucose was due to a reduction in meal size and meal frequency, and duodenal or portal vein infusions of glucose reproduced this effect [94]. It is also interesting in this context that the putative satiety hormone glucagon (see below), which is released during meals [196] and which also affects meal size [213, 214], may induce satiety by activation of a hepatic mechanism [196, 213, 215] and may actually account for some of the prandial metabolic changes described above. Therefore, the temporal pattern of metabolic changes during eating may be crucial for any effect of metabolic factors on meal size.

To summarize, the data presented so far indicate that experimentally induced changes in the hepatic oxidation of different metabolites can affect eating behavior through signals that are conveyed to the brain via vagal afferents. Moreover, comparable changes in liver metabolism seem to occur in relation to meal-taking and may well provide the physiological basis for the hepatic contribution to the metabolic control of eating. This raises the question of how the metabolic message is transformed into a neural signal. The following section will address this question.

C. Coding Mechanism of Hepatic Metabolic Sensors

Although the existence of hepatic glucosensors was postulated almost 30 years ago, the morphological substrate of these glucosensors or any other putative metabolic sensor in the liver has not yet been identified. Nevertheless, there is good reason to believe that both liver nerves and hepatocytes have sensory functions (see below) and that the utilization of glucose and other metabolic fuels is linked to the spike frequency of afferent nerves through oxidative phosphorylation and sodium pump activity [162, 216].

The classical models describing the regulation of oxidative phosphorylation in the cell mainly focussed on the important role of ATP hydrolysis products as controlling factors. However, recent data indicate that simple kinetic feedback of ATP hydrolysis products is not an adequate explanation for the complex regulation of oxidative phosphorylation [217]. In addition to ATP hydrolysis products, the delivery of reducing equivalents to the mitochondrial respiratory chain and the supply of oxygen are also considered important factors in the regulation of oxidative phosphorylation. The delivery of reducing equivalents and oxygen seems to effectively increase the rate of ATP synthesis and, hence, ATP turnover at a given concentration of ADP and P_i [217]. Blood flow appears to be an important contributor to tissue oxygen consumption and metabolic rate [218]. Thus, it appears reasonable to assume that the often observed prandial increase in splanchnic blood flow [e.g. 219] complements the prandial increases in portal metabolite levels described above and increases the rate of oxidative phosphorylation in hepatocytes and hepatic nerve endings, which finally convey the metabolic message to the brain.

The sodium pump directly or indirectly controls many essential cellular characteristics such as cell volume, heat production, intracellular pH and membrane potential and is considered a major consumer of ATP in the body [e.g. 220, 221]. Although ATP synthesis is tightly coupled to ATP hydrolysis under most physiological conditions [217, 222], dramatic transient fluctuations in the cytosolic $ATP/ADP \times P_i$ ratio have been found in transition from high to low activity or vice versa [222]. Such fluctuations seem to occur in the liver also in response to eating. Thus, the cytosolic ATP/ADP ratio and sodium pump-dependent respiration in the liver are much higher in ad libitum-fed animals than in starved animals [220, 223]. Refeeding rapidly restores the high cytosolic ATP/ADP ratio [223]. Further, fatty acids stimulate respiration and ATP synthesis in isolated hepatocytes, and a large part of this ATP is utilized by the sodium pump as seen by the fact that ouabain partially (50–60%) inhibited the stimulatory effect of fatty acids on respiration [224]. It is also interesting that the ATP supply for the sodium pump of hepatocytes is limited due to the relative large distance of the cell membrane from the mitochondria [225]. Thus, hepatocyte sodium pump activity declines dramatically when cytosolic ATP concentration decreases by about 50% [225]. This high sensitivity of the membranal sodium pump to fluctuations in the supply of ATP appears well suited to translate fluctuations of the metabolic rate and of oxidative phosphorylation into changes in hepatocyte membrane potential. Interestingly, ATP derived from glycolysis predominantly feeds the sodium pump in most tissues [225]. As mentioned before, glycolysis seems to prevail over gluconeogenesis in the liver during eating and for a considerable time thereafter [190]. Therefore, it is tempting to speculate that under physiological conditions the switch from glycolysis (=ATP production in cytosol and mitochondria) to gluconeogenesis (=ATP consumption in cytosol) may limit the supply of ATP for the sodium pump and affect hepatocyte membrane potential. The ensuing decrease in hepatocyte membrane potential might increase the spike frequency of adjacent afferent nerves, thus contributing to the development of hunger. Indirect support for a role of the cytosolic ATP/ADP ratio in the control of hunger and satiety is derived from studies using the fructose analogue 2,5-anhydro-D-mannitol, which has recently been shown to increase food intake after intraperitoneal injection in rats [226]. The exact mechanism whereby 2,5-anhydro-D-mannitol stimulates eating has not yet been identified. However, a change in the cytosolic ATP/ADP ratio may well be involved because in addition to inhibiting gluconeogenesis and glycogenolysis [227], 2,5-anhydro-D-mannitol has been shown to decrease the cytosolic ATP/ADP ratio in isolated hepatocytes [228]. More direct evidence for a role of sodium pump activity and membrane potential in the control of eating is derived from studies in which ouabain stimulated eating after intraperitoneal injection in rats [216]. The hyperphagic effect of ouabain was blocked by hepatic vagotomy [216] (fig. 6), but not by peripheral anticholinergic blockade with atropin [216]. Therefore, hepatic afferents seem to be involved in the hyperphagic effect of ouabain.

Hepatocyte membrane potential responds to a variety of stimuli [229]. Fasting for 24 h has been shown to lower hepatocyte membrane potential in situ [229, 230]. In addition, it has been shown in the perfused liver that the liver cell membrane can be hyperpolarized by palmitate, pyruvate, lactate, alanine and fructose [231]. The hyperpolarizing effect of pyruvate

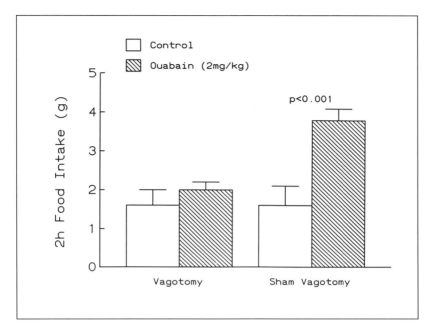

Fig. 6. Effect of hepatic branch vagotomy on the eating response to intraperitoneally injected ouabain, which inhibits the sodium pump. From Langhans and Scharrer [181].

on the liver cell membrane was reversed by ouabain and seems therefore to depend on the sodium pump [231]. In another study, pyruvate also decreased the spike frequency of hepatic vagal afferents [162]. These data are compatible with the assumption that the oxidation of fatty acids and other fuels is sensed by liver cells, which are in contact with afferents. A causal relationship between fluctuations in hepatocyte membrane potential and changes in the spike frequency of hepatic afferent nerves remains to be demonstrated. However, such a relationship is conceivable because nerve fibers seem to have intimate contact with unmyelinated, predominantly afferent fibers [158].

As already mentioned, intraportal glucose decreased the spike frequency of hepatic vagal afferents [160–162], and this effect of glucose was blocked by ouabain, suggesting that glucose utilization and the sodium pump are involved. However, discrepant findings exist concerning the effect of glucose on hepatocyte membrane potential. Dambach and Friedmann [23] failed to detect a hyperpolarizing effect of glucose in the perfused rat liver. Yet, recent in vitro data indicate that glucose has a hyperpolarizing effect on hepatocytes which is independent of intracellular glucose utilization [232]. This is compatible with previous data from in vivo experiments, in which intraintestinal administration of glucose led to a clear hyperpolarization of hepatocytes in rats [230]. Since the unmetabolizable glucose analogue 3-O-methylglucose also hyperpolarized the liver cells [230], hepatocellular glucose metabolism does not seem to be involved. It is noteworthy in this context, that alanine also inhibits eating after parenteral administration in rats [unpubl. results] and increases hepatocyte membrane potential [229, 231]. Similar to the hyperpolarizing effect of glucose, the hyperpolarizing effect of alanine does not seem to be exclusively coupled to a stimulation of sodium pump activity [229, 233]. Rather, some evidence implicates a volume-regulatory efflux of K^+ in the hyperpolarizing effect of alanine [229]. In summary, if hepatocytes act as metabolic sensors, as seems likely, metabolites may affect hepatocyte membrane potential through mitochondrial oxidation, cytosolic ATP and sodium pump activity (e.g. pyruvate, fatty acids) or through potassium efflux due to cell swelling and the ensuing opening of potassium channels (glucose, alanine) [229, 232]. The latter possibility may also apply to glucose that is derived from hepatic glycogenolysis. It remains to be clarified whether intracellular glucose oxidation also affects hepatocyte membrane potential under some conditions, for instance when glycolysis in hepatocytes is stimulated by insulin [234, 235]. Furthermore, hepatic afferents probably function as the main hepatic glucosensors, as originally suggested by Niijima [160, 162]. This seems reasonable because glucose is the preferred metabolic fuel of peripheral nerves, while hepatocytes are able to oxidize almost all other fuels and in particular fatty acids [184]. The proposed dual role of hepatic afferents as hypothetical glucosensors, which monitor their own rate of glucose utilization, and as an afferent link that conveys information of hepatocyte oxidative metabolism to the brain is also compatible with reports of synergistic interactions between the effects of glucose and other metabolites on eating [39, 120].

Recently, hepatic amino acid sensors responding to alanine, arginine and leucine have been identified [236, 237]. Hepatic vagal afferents responded to intraportal amino acids with an increase in the firing rate [237]. The mechanism of this increase and the possible interactions between amino acid sensitive and glucosensitive neurons remain to be clarified. Nevertheless, the existence of hepatic amino acid sensors is inter-

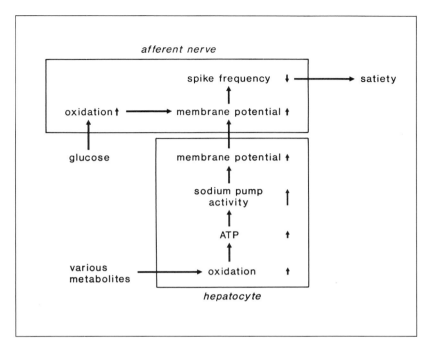

Fig. 7. Hypothetical coding mechanism for hepatic metabolic sensors involved in eating control. See text for further details.

esting because intraportal or intraperitoneal administration of amino acids has been shown to inhibit eating [e.g. 132].

In summary, utilization of various metabolic fuels in hepatocytes and the ensuing changes in hepatocyte membrane potential may combine with the effects of glucose utilization within hepatic nerves and determine the spike frequency in hepatic afferents that is ultimately integrated in the brain circuitary controlling eating behavior. Main features of this idea of a hepatic sensory function in the metabolic control of eating are shown in figure 7.

In order to fully comprehend the possible function of hepatic metabolic sensors in the control of eating, it must be kept in mind that the hepatic afferents that seem to be involved are also part of a reflex network that plays an important role in blood glucose regulation [162]. For instance, portal vein hyperglycemia inhibited adrenal efferent nerve activity and facilitated pancreatic and hepatic efferent nerve activity [162, 238]. This presumably blocks catecholamine release and results in increased insulin secretion from the pancreas and increased glycogen synthesis in the liver. Sectioning of the hepatic branch of the vagus blocked the autonomic responses to portal vein hyperglycemia, and electrical stimulation of hepatic afferents, that supposedly mimics a decrease in portal vein blood glucose level, had the opposite effect to glucose [162, 238]. Moreover, hepatic glucosensitive afferents might form the afferent limb of a hepato-gastric reflex [239], which may be involved in gastric secretion and motil-ity. Besides glucose, intraportal L-arginine also affected the firing rate in efferent pancreatic nerves [236]. These results suggest that changes in por-tal vein glucose and amino acid concentrations trigger autonomic compen-satory mechanisms through activation of hepatic sensors. Whether other metabolic fuels elicit similar compensatory autonomic responses remains to be clarified.

In addition to producing ATP through oxidative phosphorylation, oxi-dation of metabolic fuels is always thermogenic. The concomitant rise in temperature might provide an additional link between stimulation of hepatic oxidative metabolism and inhibition of eating. The hepatic vagus nerve has been shown to contain thermosensitive fibers, which respond to slight increases in liver temperature with an increase in spike frequency [240, 241]. Moreover, artificial heating of rat liver within physiological limits decreased food intake [242] and this effect appeared to be mediated by afferent nerves [243]. Stimulation of hepatic thermogenesis was impli-cated in the thermic effect of meals early on [244]. In a more recent study in which liver temperature of ad libitum-fed rats was continuously record-ed, liver temperature invariably increased to about 39 °C during meals taken in the dark phase of the lighting cycle [245]. The meal-end liver temperatures represented the highest values recorded. In a recent study in man, a strong positive correlation was observed between the deceleration in the cumulative food intake curve during meals and a temperature increase of the skin in proximity to the liver. Skin temperature of other locations and core temperature did not change during and after meals [246]. It is interesting in this context that fatty acid oxidation in liver cell mitochondria seems to be particularly thermogenic [247–249]. Whereas the oxidation of acetyl-CoA to CO_2 is considered to be preferentially cou-pled to ATP synthesis, oxidation of long- and medium-chain fatty acids to acetyl-CoA is obligatorily associated with a highly thermogenic process called redox cycling or reversed electron transfer between cytosol and mitochondria [247, 248]. Up to 30% of the bond energy of fatty acid molecules can be expected to dissipate as heat through this process [247]. While oxidation of long-chain fatty acids decreases during eating, hepatic oxidation of medium-chain fatty acids derived from dietary lipids may actually increase during fatty meals. Being released during digestion, medium-chain fatty acids are easily transferred to the liver through the portal circulation and are rapidly oxidized in the mitochondrial fatty acid oxidation system [184].

According to the results presented in this section, prandial hepatic thermogenesis may be sensed by thermosensitive afferents and combine with other ouabain-sensitive metabolic signals in the production and maintenance of satiety under certain conditions. Hepatic oxidation of fatty acids may be a particularly important contributor to this effect due to its high thermogenic capacity. It is rather unlikely, however, that hepatic thermogenesis plays a major role in the metabolic control of eating under physiological conditions. Yet, it is conceivable that hepatic thermogenesis would limit eating in order to protect the liver from overheating under certain conditions.

D. Effects of Liver Denervation on Eating

If physiological hunger and satiety signals are derived from hepatic oxidative metabolism and conveyed to the brain through afferent nerves, one might expect that liver denervation by itself would affect eating. Contrary to this assumption, partial or total liver denervation generally failed to substantially affect cumulative food intake [250–253], meal patterns [251, 252] or body weight gain [250, 252] in animals. Even orthotopic liver transplantation seems to have no profound effect on eating [254, 255]. However, these data do not exclude a role of the liver in control of food intake. As mentioned previously (this section, part B), interpretation of hepatic vagotomy results is generally limited because fibers of the hepatic vagus do not exclusively innervate the liver and do not provide the only afferent connection between liver and brain [176]. Furthermore, even if liver denervations are actually complete, the findings merely demonstrate that the liver and its innervation are not necessary for an adequate control of food intake under normal conditions. The same applies to signals arising from the gastrointestinal tract [e.g. 256, 257]. The general failure to detect necessary signals for food intake control is presumably due to the redundancy of the whole control system. For example, with the elimination of metabolic cues from the liver, metabolic sensors located in the brain (see sector VII) could gain importance. Interestingly, rates of hypothalamic fatty acid and glucose oxidation follow a similar pattern as seen in the liver of animals in a catabolic or anabolic state [258, 259]. Furthermore, internal satiety signals may have to be calibrated by learning [260, 261], i.e. they may have to be associated with familiar oral cues to be effective. Therefore, if metabolic signals from the liver provide one of the unconditioned stimuli in the control of eating, which seems likely, normal eating behavior after liver denervation may be maintained by the remaining unconditioned stimuli and by gustatory conditioned stimuli as long as the taste of the diet is familiar. Finally, it should also be mentioned that several subtle changes in eating behavior in response to liver denervation have been observed: Friedman and Sawchenko [253] reported that hepatic branch vagotomy in male rats produced a modest body weight gain and a hyperphagia, which was due to an increase in daytime food consumption. The effect on food intake was observed immediately after nerve section, which was produced in unrestrained, unanesthetized animals by pulling on a previously implanted suture [253]. Recently, Louis-Sylvestre et al. [255] also found that daytime food intake was increased while nighttime food intake was decreased between 15 and 25 days after orthotopic liver transplantation in rats. Kraly et al. [257] reported that rats kept on a sweet milk diet became hyperphagic and gained body weight at a greater than normal rate after hepatic branch vagotomy. In another study, hepatic branch vagotomized rats ate transiently more than sham-vagotomized rats when a diet with a novel and strong taste was presented immediately after surgery [262]. This is compatible with other findings indicating that an unconditioned stimulus for calorie-based learning is in fact transduced in the liver and conveyed by the hepatic branch of the vagus [91, 263]. Finally, in a study performed in our laboratory, hepatic branch vagotomized rats increased food intake somewhat more than sham-vagotomized rats when switched from a high to a low calorie diet [264] and more recent experiments showed that hepatic branch vagotomy in rats fed a high fat diet increased food intake during the early dark phase of the lighting cycle [Delprete and Scharrer, unpubl. data].

In summary, liver denervation studies clearly demonstrate that information derived from hepatic metabolism, like other peripheral signals, is not necessary for an adequate control of food intake. Nevertheless, the modest and occasionally transient effects of liver denervation on food intake under some conditions support the hypothesis that the liver is involved in the metabolic control of eating.

VI. Role of Pancreatic Hormones in the Metabolic Control of Eating

Hormones are major factors which govern the selection of respiratory fuels and the rate of respiration. Hormones involved in this process include insulin, glucagon, growth hormone, thyroid hormones, catecholamines and cortisol. The rate of release of most of these hormones is directly or indirectly affected by nutrient availability and at least some of them may also be involved in the metabolic control of eating. In this section we will focus on insulin and glucagon, which appear to be specifically involved in the short-term physiological control of eating. The effects of both hormones on food intake and metabolism and the putative relationships between these behavioral and metabolic effects have been intensively investigated.

A. Insulin

The sight, taste, and smell of food trigger an anticipatory, preabsorptive insulin release during the early phase of a meal in man [265–267] and animals [268–271]. This 'cephalic' phase of insulin release may serve to prepare the organism for the nutrients arriving from the gut. It seems to be mediated through vagal efferents, because it is blocked by vagotomy [269] and attenuated by peripheral atropinization [272]. Later into a meal, a second and greater release of insulin is prompted by gastrointestinal hormones (e.g. cholecystokinin, gastrin, gastric-inhibitory polypeptide, secretin) [see 273], which are released from chromaffin cells upon the presence of nutrients in the gastrointestinal tract. Finally, as long as blood glucose levels are high during and after a carbohydrate-containing meal, the meal-induced insulin release is complemented by the stimulating effect of glucose and, to a lesser extent, also amino acids on insulin release from the pancreas [274, 275].

Insulin is essential for the entry and metabolism of ingested nutrients in most tissues and has therefore been implicated in the metabolic control of food intake from early on. Administration of large, hypoglycemic doses of insulin generally elicits eating in man [276] and most animal species [e.g. 6, 277–280]. Stimulation of eating by insulin treatment is presumably triggered mainly by the acute cellular glucoprivation that follows the initial phase of increased glucose utilization after insulin administration. Although blood glucose falls within a few minutes, eating often does not commence until about 30–60 min after insulin injection [281]. Acute insulin treatment produced hypoglycemia and stimulated ingestion of a sucrose solution in decerebrate rats as well as control rats [282]. Interestingly, however, ablation of the area postrema and of the adjacent caudomedial portions of the nucleus of the solitary tract attenuated the short-term eating response to insulin [283]. Therefore, neural systems of the caudal brainstem seem to be involved in the acute eating response to exogenous insulin. Metabolic sensors in the periphery have also been implicated because infusion of fructose, which does not cross the blood-brain barrier [284], after insulin administration also attenuated insulin-induced hyperphagia [285], although somewhat less than glucose [137]. Fructose did not increase blood glucose levels under the same conditions [285], indicating that the inhibitory effect of fructose on insulin-induced hyperphagia was not due to a conversion of fructose to glucose. Interestingly, insulin-induced hyperphagia cannot be completely attributed to secondary glucoprivation because high doses of insulin have also been shown to increase food intake when hypoglycemia was prevented by the concomitant infusion of glucose [286]. The mechanism of insulin's hyperphagic effect under these conditions is unknown. In contrast to the stimulation of eating in response to acute insulin administration [283], the hyperphagic response to chronic insulin administration is not affected by lesions of the area postrema and the nucleus of the solitary tract [287]. This suggests that the injection of high, hypoglycemic doses of insulin and chronic insulin administration of moderate doses stimulate eating through different neural substrates.

Hyperinsulinemia and hyperphagia without hypoglycemia can be found in genetic and experimentally induced obesity in laboratory animals as well in human obesity [288–291]. Hyperinsulinemia in obesity is usually accompanied by insulin resistance [292, 293] and has often been implicated in the initial hyperphagia observed in different types of obesity [291–295]. The hyperphagia and body weight gain that has sometimes been observed with chronic peripheral administation of low insulin doses [296] is compatible with this assumption. Circadian fluctuations of plasma insulin levels and insulin-induced lipogenesis have also been implicated in the circadian fluctuations of food intake in man and rat [26, 100, 108]. Indirect support for this assumption is derived from the finding that both insulin-induced hyperphagia and the nocturnal hyperphagia in the rat are characterized by an increase in meal frequency [296, 297]. The hyperphagia that is normally observed during chronic experimental diabetes [298, 299] does not necessarily argue against a physiological hunger effect of insulin because the hyperphagia that accompanies diabetes is due to the inability to utilize glucose [300]. In fact, an initial suppression of food intake has been observed in the early phase of streptozotocin-induced diabetes [298]. In summary, large doses of exogenous insulin clearly stimulate eating. Some data suggest that endogenous insulin might enhance eating, at least under some conditions, but the evidence for a physiological hunger effect of endogenous insulin is inconclusive.

On the other hand, long-term intraperitoneal infusions [301, 302] or intravenous injections [303–305] of small amounts of insulin have been reported to reduce rather than enhance food intake in rat [301, 302], pig [303] and sheep [304, 305]. Insulin reduced food intake by suppressing meal size, suggesting that modest elevations of plasma insulin (as would occur in response to eating) participate in the induction of satiety [301]. In line with this idea, low doses of the sulfonylurea tolbutamide produced a mild increase in plasma insulin level and decreased food intake, whereas higher doses of tolbutamide had the opposite effects [306]. Furthermore, a stimulation of eating has been observed in the rat after subdiabetogenic doses of streptozotozin [307]. Sham feeding was also diminished by intraperitoneal injection of insulin [308]. Whether the satiety effect of peripherally-administered insulin is mediated by a peripheral metabolic effect of insulin or by direct action of insulin on the brain is not known. Insulin might act rapidly on neurons of brain areas that lack a functioning blood-brain barrier. This applies to the periventricular organs, and includes the area postrema, and hypothalamic areas where insulin receptors have been found [309]. Another important and possibly fast passage for insulin from blood to brain sites that are involved in food intake regulation seems to be via specific insulin receptors in the brain capillary endothelial walls [310]. Evidence for a central satiety effect of insulin is derived from several reports of hypophagia and body weight loss in response to intracerebroventricular infusions of insulin [311–313]. Intrahypothalamic administration of insulin also reduced food intake [77, 314] and intrahypothalamic injection of antibodies to insulin increased food intake in rats [315]. Whether insulin's central satiety effect depends on glucose is not definitely known, but the ineffectiveness of intracerebroventricularly infused insulin to depress food intake in rats fed a low carbohydrate, high fat diet [140] suggests that glucose might be involved. In addition, Giza and Scott [316, 317] have shown that taste responsiveness to glucose and fructose was significantly reduced following intravenous infusion of insulin or glucose in rats. Thus, a diminished hedonic appeal of food might also contribute to insulin-induced satiety. Finally, insulin may also induce satiety through a peripheral metabolic effect. This assumption is compatible with earlier reports of a rapidly enhanced satiety effect of ingested nutrients [318] or infused glucose [35] by insulin. At physiological blood glucose concentrations, insulin stimulates hepatic glucose uptake [319, 320] and blocks hepatic glucose release [235]. Interestingly, insulin potently stimulates both hepatic glycogen synthesis [319] and glycolysis [234, 235]. One major site of intracellular insulin action seems to be the mitochondrial Krebs cycle [321]. Therefore, since both insulin release from the pancreas and hepatic glycolysis increase in response to eating, it is conceivable that insulin inhibits eating by promoting hepatic glucose utilization. Clearly however, further experiments are necessary to test this idea.

B. Glucagon

The evidence for a physiological satiety effect of glucagon is compelling and has been discussed in a recent comprehensive review by Geary [322]. Therefore, only some data which are pertinent for the possible role of glucagon in the metabolic control of eating will be presented below.

Textbook knowledge holds that the plasma level of glucagon increases during fasting and decreases in response to eating, when glucose, which inhibits glucagon secretion, becomes more readily available. Nevertheless, plasma glucagon has often been shown to increase during mixed nutrients' meals in various species [196, 270, 323] including man [274, 324]. The prandial increase in plasma glucagon level commences within the first minute of eating [270]. Thus, it seems to be triggered mainly by neural and endocrine stimuli and is probably influenced less by substrates absorbed from the intestine. Activation of the sympathetic nervous system seems to be a major stimulus for pancreatic glucagon release [325]. If prandially released endogenous glucagon contributes to satiety, antagonism of glucagon during meals should increase meal size. This has in fact been shown in rats. When glucagon antibodies sufficient to neutralize 5 ng pancreatic glucagon were injected just before the first nocturnal meal after 12 h of food deprivation, meal size and meal duration increased by 63 and 74%, respectively [214]. Recently, the remotely controlled intraportal infusion of a lower dose of glucagon antibodies (sufficient to neutralize 1.5 ng glucagon) was shown to increase the size of spontaneous meals in undisturbed, non-food-deprived rats by 73% [326]. These findings clearly implicate prandially released endogenous glucagon in normal satiety. In line with this hypothesis, administration of pancreatic glucagon has often been shown to reduce food intake in man [322, 327, 328] and animals [213, 305, 329–

331]. The inhibition of eating by glucagon is generally rapid but transient, suggesting that glucagon affects primarily meal size. LeSauter and Geary [332] recently demonstrated that remotely controlled intraportal glucagon infusions reduced the size and duration of spontaneous nocturnal meals in undisturbed rats. This finding extends the previous demonstrations of the inhibition of eating by glucagon in scheduled test meals [213, 333], and is particularly relevant to glucagon's satiety effect from an ecological point of view. The glucagon doses required for an inhibition of eating appear large compared to endogenous glucagon levels observed during eating [see 322]. However, there is evidence that very low doses of glucagon that presumably do not raise plasma glucagon levels above the physiological range can also inhibit eating [305, 322]. On the whole, there is little doubt that glucagon acts as a satiety hormone.

The liver is glucagon's primary metabolic target organ and may also be involved in glucagon's satiety effect. Interestingly, simultaneous measurements of portal vein and hepatic vein plasma glucagon levels in rats during meals instigated by mild food deprivation indicate that prandially secreted glucagon is extracted mainly by the liver [196]. Similar observations have been made in humans [324]. Further evidence for a hepatic origin of glucagon's satiety effect is derived from denervation studies. Exogenous glucagon failed to decrease food intake after subdiaphragmatic vagotomy [334, 335], and in some [215, 336, 337], although not all [e.g. 338], experiments after hepatic branch vagotomy. In contrast, abdominal vagotomy that spared only the hepatic branch of the vagus did not affect glucagon's satiety effect [215]. Likewise, dissection of hepatic vagal fibers that do not follow the hepatic branch of the vagus also failed to affect glucagon satiety [336]. Collectively these data suggest that the hepatic branch of the vagus is a necessary and sufficient contribution of the liver's innervation to glucagon-induced satiety. Furthermore, the critical lesion that blocks glucagon-induced satiety appears to be sensory because capsaicin, which is toxic to small diameter afferents, and selective lesion of the terminal fields of hepatic vagal afferents in the nucleus of the solitary tract blocked the satiety effect of glucagon [339, 340]. In summary, these findings strongly suggest that glucagon's satiety effect originates in the liver and is mediated by hepatic vagal afferents.

It was originally proposed that glucagon signals satiety by stimulating hepatic glucose production [327]. Glucagon-induced satiety in fact sometimes appeared to be linked to its glycogenolytic and hyperglycemic effects. Thus, glucagon inhibited eating less potently in food-deprived rats, in which liver glycogen was depleted [341]. Furthermore, intraperitoneal injection of glucagon at meal onset, which reduced meal size and meal duration, enhanced the prandial hepatic glycogenolysis and hepatic glucose release [213], whereas injection of glucagon antibodies had the opposite effect [214]. More recently, Ritter and her colleagues [342] provided interesting evidence that glucagon-induced satiety is based on a hepatic glucosensitive mechanism. They showed that intraportal injection of sub-diabetogenic doses of alloxan abolished glucagon-induced satiety, yet intrajugular injection of the same alloxan dose had no effect on glucagon-induced satiety [342]. Alloxan is toxic to glucosensitive cells and its effect seems to be limited on the liver under the conditions tested [342]. This suggests that glucosensitive cells in the liver, possibly vagal afferents (see section V), are necessary for glucagon-induced satiety. Other findings do not support the assumption that glucagon-induced satiety depends on glucagon's glycogenolytic or hyperglycemic effects. Indeed, several dissociations regarding the effects of glucagon on food intake and hepatic glucose production have been reported [333, 341, 343–345]. Actually, glucagon's potency to inhibit eating sometimes appeared to be inversely related to hepatic glucose release [343, 345]. Consequently, glucagon-induced hepatic glycogenolysis or hepatic glucose release is presumably not directly involved in glucagon-induced satiety. Yet, this does not necessarily exclude the possibility that hepatic glucosensors are involved in glucagon-induced satiety because hepatic glucosensors are supposed to monitor glucose utilization rather than glucose release from hepatocytes. It is also interesting that glucagon-induced glycogenolysis is followed by a short stimulation of hepatic glycolysis [346, 347], which might be related to a Ca^{2+}-dependent inhibition of hepatic glucose-6-phosphatase activity [348] and to a stimulation of insulin release from the pancreas by glucagon. Weick and Ritter [349] suggested that insulin-dependent increases in hepatic glucose utilization may cause glucagon satiety because the inhibitory effect of glucagon on meal size was related to plasma insulin changes, but not to plasma glucose changes. However, the different effectiveness of glucagon and insulin to suppress sham feeding [308, 343] is difficult to explain with this hypothesis. In short, the question of whether a hepatic glucosensitive mechanism contributes to glucagon-induced satiety or not remains open to discussion.

Besides stimulating hepatic glycogenolysis and gluconeogenesis, glucagon activates hepatic oxidative metabolism [190, 346, 350], in particular fatty acid oxidation and ketogenesis [190, 351], and increases hepatocyte membrane potential [352–354]. The stimulating effect of glucagon on hepatic mitochondrial respiration was originally observed in isolated mitochondria [350], but has subsequently been observed in situ as well [355]. Glucagon may facilitate mitochondrial respiration through an inhibition of the Ca^{2+} pump of the cell membrane [356, 357]. Interestingly, the 19–29 amino acid fragment of pancreatic glucagon is considered the true inhibitor of the hepatic Ca^{2+} pump [358] and the same fragment is apparently responsible for glucagon's satiety effect because the 1–21 amino acid fragment of glucagon did not affect meal size when injected intraperitoneally in rats [359]. Stimulation of hepatic mitochondrial respiration or fatty acid oxidation by glucagon may finally lead to hepatocyte hyperpolarization and, hence, contribute to glucagon's satiety effect at least under some conditions (see section V).

Alternatively, glucagon may hyperpolarize hepatocytes directly by stimulating sodium pump activity or by increasing membrane K^+ permeability [352, 353, 360]. Finally, glucagon has also been shown to decrease the spike frequency in hepatic vagal afferents both in vivo and in isolated perfused tissue [162].

In summary, the evidence for a specific satiety effect of glucagon is compelling. Although there are some open questions about the exact mechanism of glucagon's satiety effect, most data indicate that glucagon-induced satiety originates in the liver and is in one way or another related to glucagon's effects on hepatic metabolism and membrane function. This coincides with the hypothesis that an increase in the hepatic utilization of metabolic fuels is linked to satiety by sodium pump, membrane potential, and spike frequency in afferent hepatic nerves.

VII. Metabolic Sensors in the Brain

Mayer [58] originally proposed that the glucostatic control of food intake would be integrated into the homeostatic mechanisms of the brain that depends almost exclusively on oxidation of glucose. Oomura and Yoshimatsu [177] first provided electrophysiological evidence for the existence of two types of neurons that respond to glucose in the brain: The so-called glucoreceptor neurons in the ventromedial hypothalamus (VMH), whose activity is increased upon systemic or direct electrophoretic application of glucose [361], and the so-called glucose-sensitive neurons in the lateral hypothalamus (LH), which are hyperpolarized and have their activity suppressed by glucose [362]. Both the VMH and the LH have long been implicated in the control of food intake and energy balance. Recently, glucose-sensitive neurons have also been indentified in the nucleus of the solitary tract (NST) [363] and in the area postrema (AP) [364].

An involvement of central glucose-sensitive neurons in hunger is suggested by findings of an increase in food intake in response to intracerebroventricular infusion of low doses of 2-DG [365, 366]. Another nonmetabolizable glucose analogue, 5-thioglucose (5-TG), seems to be even more effective than 2-DG in stimulating food intake after intracerebroventricular administration [367]. Although intracerebroventricular administrations of glucose antimetabolites reliably stimulated eating, attempts to elicit eating with intrahypothalamic 2-DG applications generally failed [365, 366], suggesting that local glucoprivation in the hypothalamus is not sufficient to initiate eating. Furthermore, recent findings suggest that glucoprivic eating requires glucose-sensitive neurons in the AP/NST region of the hindbrain [73]. Thus, when the aqueduct between the third and the fourth cerebral ventricles was obstructed, microinjections of 5-TG into the fourth ventricle stimulated eating effectively, whereas equivalent injections into the lateral ventricle did not. Moreover, the eating response to intracerebroventricular and systemic glucoprivation was impaired by administration of low doses of alloxan, a toxin that selectively destroys glucose-sensitive cells, into the fourth ventricle [368]. These findings suggest that central glucose-sensitive neurons are also involved in the hyperphagic effect of peripheral glucoprivation. Finally, ascending noradrenergic fibers that originate in the hindbrain and project to the hypothalamus also seem to be important for glucoprivic eating [369].

Although intrahypothalamic injections of glucose predominantly failed to affect eating [see 369], glucose given intracerebroventricularly decreased food intake at least in some studies [370–372]. In one study [371], intracerebroventricularly injected glucose caused an immediate dose-dependent reduction of meal size without affecting meal frequency. The authors related the satiety effect of glucose to an activation of ventromedial hypothalamic glucoreceptor neurons [371] because glucose supposedly activates these neurons [361] and because electrical stimulation of the neurons has been shown to terminate ongoing eating [373]. This interpretation is compatible with the original concept of a ventromedial hypothalamic satiety and a lateral hypothalamic eating system [374]. Nevertheless, the interactions between circulating glucose, brain glucoreceptors and eating seem to be more complex than originally thought. The glucoreceptor neurons in the VMH can also activate the peripheral sympathetic nervous system [375]. Furthermore, it has been recently shown that a decrease in circulating glucose is associated with a decrease in α_2-adrenergic receptor binding and a concomitant increase in norepinephrine turnover in the hypothalamus, in particular in the paraventricular nucleus (PVN) [376–379]. This is interesting because central injection studies demonstrate a potent stimulatory effect of norepinephrine on eating, which is obviously mediated by α_2-adrenergic receptors located in the hypothalamic PVN [see 379]. Further studies suggest that the α_2-adrenergic stimulation of eating is particularly important at the onset of the active cycle, influenced by circulating cortisol, and directed specifically to carbohydrates [see 379]. As already mentioned (see section IV), a transient decline in the blood glucose level precedes and seems to be causally related to meal onset in freely feeding rats [74, 76]. Thus, an increase in hypothalamic paraventricular noradrenergic activity could well contribute to the effect of the premeal decline in blood glucose on meal initiation. Interestingly, intraparaventricular norepinephrine injections that stimulate eating have recently been shown to reduce ^{14}C-2-DG-uptake in several brain areas caudal of the hypothalamus, including the NST [380]. This is in line with recent proposals of a paraventricular-hindbrain feeding system, and suggests that this system includes hindbrain glucose-sensitive neurons. In addition to causing changes in paraventricular norepinephrine turnover, slight and transient changes in blood glucose levels have also been shown to affect the spike frequency of cells in the lateral hypothalamic area [381]. A majority of the responsive cells displayed either an activation during hypoglycemia or a depression during hyperglycemia [381], which is in line with the characateristics of the lateral hypothalamic glucose-sensitive cells identified by Oomura et al. [362]. More recently, however, the same group observed that separate lateral hypothalamic neurons responded to local or systemic glucose administration [382]. Therefore, activation of a considerable number of the glucose-sensitive, lateral hypothalamic neurons obviously involves an indirect afferent pathway that conveys information from other central and peripheral sensors, and possibly hepatic glucose sensors. Monoamines do not appear to be essential in the putative glucose-sensitive, lateral hypothalamic eating system, because 2-DG produced no clear changes in the monoamine content of the lateral hypothalamic area [383]. However, glucose-sensitive neurons in the LH can inhibit sympathetic firing rate [375], which is in line with the hypothesis that the activation state of the sympathetic nervous system is intimately connected with the metabolic control of eating [2].

Several lines of evidence suggest that γ-aminobutyric acid (GABA) in addition to norepinephrine provides a functional intermediary link between brain glucose metabolism and eating. Thus, hypothalamic GABA levels seem to be related to glucose utilization, with an increase in ventromedial hypothalamic GABA synthesis and a decrease in lateral hypothalamic GABA synthesis in response to hypo- and hyperglycemia, respectively [384]. Glucoprivation induced by peripheral or central administration of 2-DG also markedly increased the activity of ventromedial hypothalamic glutamate decarboxylase, the rate-limiting enzyme in GABA synthesis [385]. GABA or the GABA agonist muscimol increased food intake when injected into the VMH, but decreased food intake after injection into the LH [386, 387]. Medial hypothalamic injections of GABA antagonists attenuate the eating response to peripheral hypoglycemia [388]. Furthermore, the glucose flux through the GABA shunt in the LH appears to parallel the degree of satiety [259]. As GABA antagonists are also effective in blocking the eating response to paraventricular norepinephrine administration [386], GABA may mediate the effect of norepinephrine on eating (see above) and play an important role in the cascade of neurotransmitters controlling eating behavior [389].

Glycerol and DHB also reduced food intake when infused intracerebroventricularly [140, 370]. This suggests that CNS neurons involved in the control of eating respond to other metabolites in addition to glucose. In line with this assumption, direct iontophoretic application of DHB has recently been shown to have a similar effect as glucose on ventromedial hypothalamic glucoreceptor and lateral hypothalamic glucose-sensitive neurons [390]. Lateral hypothalamic glucose-sensitive neurons have also been shown to respond to microapplications of fatty acids [391]. However, this seems to be an indirect effect related to an inhibition of glucose utilization by fatty acids because the cells which are inhibited by glucose (see above) were facilitated by fatty acids [391]. Nevertheless, other data indicate that neurons in different brain areas are able to oxidize fatty acids [see 116]. In a series of experiments, Kasser et al. [258, 392] demonstrated that in vivo palmitate uptake and in vitro palmitate oxidation in the LH were inversely related to the level of satiety. More recent results from the same group indicated, however, that lateral hypothalamic fatty acid oxidation is less responsive to acute energy intake. Rather, it seems to be proportional to the level of peripheral fat storage and may therefore be involved in the long-term regulation of food intake and energy balance by body fat [393]. In contrast, glucose oxidation in the LH and in the brainstem seemed to respond to concurrent energy intake [393] and may therefore be more important for the short-term control of hunger and satiety from meal to meal.

In summary, some data indicate that physiological fluctuations of glucose utilization related to meal-taking are monitored by glucose-sensitive neurons in various brain areas. Whether this contributes to the metabolic control of eating as originally suggested by Mayer [58] remains to be proven. Furthermore, glucose availability may be linked to eating behavior, at least in part by monoaminergic pathways. As yet, the evidence for a similar function of central neurons that are sensitive to the utilization of other fuels in addition to glucose appears to be rather weak.

VIII. Metabolic Cues and Taste Perception

Common experience confirms the results of psychophysiological experiments, which demonstrate that foods are more palatable with hunger and become less palatable with the development of satiety [394]. Rolls et al. [395] found that the pleasantness of the taste of foods eaten to satiety decreased more than it did for foods that had not been eaten. This hedonic change of sensory inputs contributes to satiety [395]. Consequently, if a variety of foods is available, the total amount consumed is more than when only one food is offered [396, 397]. Since the pleasantness of food also declines when it is only tasted, but not ingested [e.g. 398], external sensory stimulation seems to play a role in the diminishing appeal of food during and after ingestion. Accordingly, this type of satiety has been termed sensory-specific satiety [395]. In addition to the effect of external sensory stimulation, the physiological outcome of ingestion apparently also affects the hedonic value placed on the taste of ingested foods [394, 399]. This change in the hedonic value, which presumably also contributes to satiety, has been related to a modulation of sensory inputs by internal signals and has been termed alliesthesia [394]. Alliesthesia is bidirectional, i.e. positive and negative shifts in taste hedonics can be observed dependent on the nutritive state [394]. Some data indicate that metabolic cues are involved in alliesthesia.

Cabanac [394] reported an inverse relationship between nutritional state as well as blood glucose levels and the hedonic value of sweet taste in humans. In another study, the pleasantness of concentrated sucrose solutions increased significantly after 2-DG infusion in man despite the concomitant hyperglycemia, suggesting the intracellular glucopenia caused the hedonic change [65]. Further evidence for a modulating effect of metabolic cues on taste perception is derived from rat studies, showing for instance that an increase in blood glucose level decreases perceived sweetness intensity [316].

It should be mentioned that sensory stimulation can affect metabolism as well, i.e. the interaction is bidirectioal. Thus, gavage feeding, which eliminates oral sensory stimulation, potently affects the utilization of absorbed nutrients [400–402]. It has also been shown that oral factors modulate blood glucose levels [77], presumably through cephalic-phase release of pancreatic hormones [77]. Recently, a modulation of vagal and splanchnic efferent activity by taste stimuli has been reported [403].

As already mentioned (see section VII), visceral vagal, splanchnic and gustatory afferents converge in caudal brainstem nuclei [174, 180, 404]. The caudal brainstem also contains neurons sensitive to glucose [177, 363] and has numerous connections with the hypothalamus and other diencephalic sites implicated in the control of eating [e.g. 178, 405, 406]. Giza and Scott reported that intravenous glucose [407], insulin [317], and glucagon [408] depressed the responsiveness of rat NST neurons to sweet taste. Although the exact mechanisms of these effects remain to be elucidated, it is tempting to speculate that the observed electrophysiological responses are related to the decrease in the pleasantness of taste associated with eating (see above).

In contrast to the effects of intravenously infused glucose and insulin on taste-evoked responses in rat NST neurons [316, 407], consumption of glucose solutions to satiety did not affect the responsiveness of NST neurons to the taste of various solutions in monkeys, even though glucose ingestion presumably increased circulating glucose and insulin [409]. Similar results were obtained with recordings from neurons in the primary gustatory cortex of the monkeys [410]. On the other hand, taste-evoked responses of neurons in the orbitofrontal cortex and in the lateral hypothalamus were suppressed by ingesting glucose to satiety. Therefore, the taste-modulating effects of circulating glucose that are present in the rat's hindbrain do not seem to be evident in primates before several more stages of central processing of the taste signal [410]. This discrepancy might also explain why humans typically report that the pleasantness but not the intensity of taste stimuli declines with satiety [395], whereas in the rat both intensity perception and pleasantness seem to decline [316].

In summary, psychophysical and electrophysiological evidence suggests that circulating glucose, and perhaps other fuels as well as hormones can modulate taste perception by affecting the taste-evoked activity of neurons in the brain. The modulation seems to occur on various levels of central processing, dependent on the species, and is implicated in the development of satiety.

IX. Concluding Remarks

As pointed out in the Introduction, the metabolic control of eating is integrated into a highly complex sequence of events that involves activation of different kinds of chemo- and mechanoreceptors in mouth, stomach and intestine, and also involves the concomitant release of gastrointestinal and pancreatic hormones. Each of these factors seems to be an important modulator of the satiating potency of other factors. In this way, interactions between all steps of the sequence ensure the characteristic redundancy of eating control and account for the successful adaptation of eating behavior to physiological needs.

Given this background, the utilization of various metabolic fuels, derived from the previous meal or from degradation of glycogen or body fat, seems to be continuously monitored by intimately connected metabolic sensors located in liver and brain. As outlined in sections V and VII, some evidence indicates that separate receptors monitor glucose utilization and the oxidation of other fuels, hence implicating that the integration of the separate signals occurs on the neural level. However, alternatively or even in addition to this possibility, some integration may occur on a biochemical level. In any case, a relative decrease in the rate of fuel utilization seems to be among the signals that trigger a meal. Under physiological conditions, this decrease may occur in response to the dwindling absorption of foodstuffs from the intestine or in response to metabolic shifts related to depletion of liver glycogen or stimulation of gluconeogenesis and lipogenesis. In addition to the putative role of metabolic cues in the maintenance of satiety between meals, a prandial increase in fuel utilization related to a stimulation of the release and utilization of stored fuels by prandially released glucagon and insulin, may complement the satiating potency of preabsorptive gastrointestinal signals and, hence, contribute to meal termination. Whatever the metabolic stimuli for hunger and satiety are, they are subject to modulation by numerous other physiological, psychological and pathological factors before the behavioral consequences become evident. Yet, despite the considerable knowledge accumulated during recent years, our understanding of the metabolic control of eating is certainly still incomplete.

References

1 Scharrer E, Langhans W: Mechanisms for the effect of body fat on food intake; in Forbes JM, Hervey GR (eds): Control of Body Fat Content. London, Smith, Gordon, Ltd., 1990, pp 63–86.

2 Bray GA: Nutrient balance: New insights into obesity. Int J Obes 1987;11(suppl 3): 83–95.

3 Cohn C, Joseph D: Influence of body weight and body fat on appetite of 'normal' lean and obese rats. Yale J Biol Med 1962;34:598–607.

4 Levitsky DA: Feeding patterns of rats in response to fasts and changes in environmental conditions. Physiol Behav 1970;5:291–300.

5 Scharrer E, Baile C, Mayer J: Effect of amino acids and protein on food intake of hyperphagic and recovered aphagic rats. Am J Physiol 1970;218:400–404.

6 Panksepp J, Pollack A, Krost K, et al: Feeding in response to repeated protamine zinc insulin injection. Physiol behav 1975;14:487–493.

7 Kennedy GC: Role of depot fat in the hypothalamic control of food intake in the rat. Proc R Soc Lond [B] 1953;140:578–592.

8 Kleiber M, Dougherty JE: the influence of environmental temperature on the utilization of food energy in baby chicks. J Gen Physiol 1934;17:701–726.

9 Van Es AJH, Vogt JE, Niessen C, et al: Human energy metabolism below, near and above energy equilibrium. Br J Nutr 1984;52:429–442.

10 Katzeff HL, Danforth E Jr: Decreased thermic effect of a mixed meal during overnutrition in human obesity. Am J Clin Nutr 1989;50:915–921.

11 Yamashita J, Hayashi S: Changes in the basal metabolic rate of a normal woman induced by short-term and long-term alterations of energy intake. J Nutr Sci Vitaminol 1989;35:371–381.

12 McCutcheon NB, Tennissen AM: Hunger and appetitive factors during total parenteral nutrition. Appetite 1989;13:129–141.

13 Auffray P, Marcilloux JC: Analyse de la séquence alimentaire du porc, du sevrage à l'état adult. Reprod Nutr Dév 1980;20:1625–1632.

14 Auffray P, Marcilloux JC: Etude de la séquence alimentaire du porc adulte. Reprod Nutr Dév 1983;23:517–524.

15 Rashotte ME, Smith JC, Austin T, et al: Twenty-four-hour free-feeding patterns of dogs eating dry food. Neurosci Biobehav Rev 1984;8:205–210.

16 Hansen BC, Jen K-LC, Kalnasy LW: Control of food intake and meal patterns in monkeys. Physiol Behav 1981;27:803–810.

17 LeMagnen J, Tallon S: Enregistrement et analyse préliminaire de la périodicité alimentaire spontanée chez le rat blanc. J Physiol (Paris) 1963;55:286–297.

18 LeMagnen J, Tallon S: La périodicité spontanée de la prise d'aliments ad libitum du rat blanc. J Physiol (Paris) 1966;58:323–349.

19 Kanarek RB: Availability and caloric density of the diet as determinants of meal patterns in the cat. Physiol Behav 1975;15:611–618.

20 Ralston SL: Control of feeding in horses. J Anim Sci 1984;59:1354–1361.

21 Chase LE, Wangsness PJ, Baumgardt BR: Feeding behavior of steers fed a complete mixed ration. J Dairy Sci 1976;59:1923–1928.

22 Baile CA: Regulation of energy balance and control of food intake; in Church DC (ed): Digestive Physiology and Nutrition of Ruminants. Corvalis Books, 1979, pp 291–320.

23 Langhans W, Senn M, Scharrer E, et al: Free feeding pattern of pygmy goats eating a pelleted diet. J Anim Physiol Anim Nutr 1988;59:160–166.

24 Hirsch E: Some determinants of intake pattern of feeding in the guinea pig. Physiol Behav 1973;11:687–704.

25 Sanderson JD, Van der Weele DA: Analysis of feeding patterns in normal and vagotomized rabbits. Physiol Behav 1975;5:1207–1209.

26 LeMagnen J, Devos M: Daily body energy balance in rats. Physiol Behav 1982;29:807–811.

27 Bernstein IL: Relationship between activity, rest, and free feeding in rats. J Comp Physiol Psychol 1975;89:253–257.

28 DeCastro JM: Meal pattern correlations: Facts and artifacts. Physiol Behav 1975;15:13–15.

29 LeMagnen J, Devos M: Meal-to-meal energy balance in rats. Physiol Behav 1984;32:39–44.

30 DeCastro JM, Balagura S: A preprandial intake pattern in weanling rats ingesting a high fat diet. Physiol Behav 1976;17:401–405.

31 Levitsky DA: Feeding conditions and intermeal relationships. Physiol Behav 1974;3:137–140.

32 Kraly FS, Cushin BJ, Smith GP: Nocturnal hyperphagia in the rat is characterized by decreased postprandial satiety. J Comp Physiol Psychol 1980;94:375–387.

33 Strubbe JH, Kexser J, Dijkstra T, et al: Interaction between circadian and caloric control of feeding behavior in the rat. Physiol Behav 1986;36:489–493.

34 DeCastro JM: Circadian rhythms of the spontaneous meal pattern, macronutrient intake, and mood of humans. Physiol Behav 1987;40:437–446.

35 Nicolaidis S, Rowland N: Metering of intravenous versus oral nutrients and regulation of energy balance. Am J Physiol 1976;231:661–668.

36 Larue-Achagiotis C, LeMagnen J: Metabolic correlates of the effects of 2-deoxy-glucose on meal size and post-meal satiety in rats. Physiol Behav 1981;26:193–196.

37 Even P, Nicolaidis S: Spontaneous and 2-DG induced metabolic changes and feeding: The ischymetric hypothesis. Brain Res Bull 1985;15:429–435.

38 Langhans W, Scharrer E: Role of fatty acid oxidation in control of meal pattern. Behav Neural Biol 1987;47:7–16.

39 Sydler T: Untersuchungen zum Wirkungsmechanismus hepatischer Sättigungssignale. Vet Med Thesis, University of Zürich, 1988.

40 Panksepp J, Ritter M: Mathematical analysis of energy regulatory patterns of normal and diabetic rats. J Comp Physiol Psychol 1975;89:1019–1028.

41 Castonguay TW, Kaiser LL, Stern JS: Meal pattern analysis: Artifacts, assumptions and implications. Brain Res Bull 1986;17:439–443.

43   Bellisle F: Human feeding behavior. Neurosci Biobehav Rev 1979;3:163–169.

43   DeCastro JM, Kreitzman SM: A microregulatory analysis of spontaneous human feeding patterns. Physiol Behav 1985;35:329–335.

44   DeCastro JM, McCormick J, Pedersen M, et al: Spontaneous human meal patterns are related to preprandial factors regardless of natural environmental constraints. Physiol Behav 1986;38:25–29.

45   Weingarten HP: Stimulus control of eating: Implications for a two-factor theory of hunger. Appetite 1985;6:387–401.

46   Birch LL, McPhee L, Sullivan S, et al: Conditioned meal initiation in young children. Appetite 1989;13:105–113.

47   Green J, Pollak CP, Smith GP: Meal size and intermeal interval in human subjects in time isolation. Physiol Behav 1987;41:141–147.

48   Bernstein IL, Zimmerman JC, Czeisler CA, et al: Meal patterns in 'free-running' humans. Physiol Behav 1981;27:621–623.

49   DeCastro JM, DeCastro ES: Spontaneous meal patterns of humans: Influence of the presence of other people. Am J Clin Nutr 1989;50:237–247.

50   DeCastro JM: Social facilitation of duration and size but not rate of the spontaneous meal intake of humans. Physiol Behav 1990;47:1129–1135.

51   Stellar E, Shrager EE: Chews and swallows and the microstructure of eating. Am J Clin Nutr 1985;42:973–982.

52   Davis JD, Smith GP: Analysis of lick rate measures the positive and negative feedback effects of carbohydrates on eating. Appetite 1988;11:229–238.

53   Clifton PG, Popplewell DA, Burton MJ: Feeding rate and meal patterns in the laboratory rat. Physiol Behav 1984;23:369–374.

54   Meyer JE, Pudel V: Experimental studies on food intake in obese and normal weight subjects. J Psychosom Res 1972;16:305–308.

55   Adams N, Ferguson J, Stunkard A, et al: The eating behavior of obese and non-obese women. Behav Res Ther 1978;16:225–232.

56   Kaplan DL: Eating styles of obese and non-obese males. Psychosom Med 1980;42:529–538.

57   Westerterp-Plantenga LS, Westerterp KR, Nicolson NA, et al: The shape of the cumulative food intake curve in humans, during basic and manipulated meals. Physiol Behav 1990;47:569–576.

58   Mayer J: Glucostatic mechanism of regulation of food intake. N Engl J Med 1953;249:13–16.

59   Brown J: Effects of 2-deoxy-*D*-glucose on carbohydrate metabolism: Review of the literature and studies in the rat. Metabolism 1962;11:1098–1112.

60   Chen M, Whistler RL: Action of 5-thio-*D*-glucose and its 1-phosphate with hexokinase and phosphoglucomutase. Arch Biochem 1975;169:392–396.

61   Smith GP, Epstein AN: Increased feeding in response to decreased glucose utilization in rat and monkey. Am J Physiol 1969;217:1083–1087.

62   Houpt TR, Hance HE: Stimulation of food intake in the rabbit and rat by inhibition of glucose metabolism with 2-deoxy-*D*-glucose. J Comp Physiol Psychol 1971;76:395–400.

63   Novin D, Van der Weele DA, Rezek M: Infusion of 2-deoxy-*D*-glucose into the hepatic portal system causes eating: Evidence for peripheral glucoreceptors. Science 1973;181:858–860.

64 Parrot RF, Baldwin BA: Effects of intracerebroventricular injections of 2-deoxy-D-glucose, D-glucose and xylose on operant feeding in pigs. Physiol Behav 1978;21: 329–331.

65 Thompson DA, Campbell RG: Hunger in humans induced by 2-deoxy-D-glucose: Glucoprivic control of taste preference and food intake. Science 1977;198:1065–1068.

66 Welle SL, Thompson DA, Campbell RG, et al: Increased hunger and thirst during glucoprivation in humans. Physiol Behav 1980;25:397–403.

67 Booth DA: Modulation of the feeding response to peripheral insulin, 2-deoxy-glucose or 3-O-methyl-glucose injection. Physiol Behav 1972;8:1069–1076.

68 Kanarek RB, Marks-Kaufman R, Ruthazer R, et al: Increased carbohydrate consumption by rats as a function of 2-deoxy-D-glucose administration. Pharmacol Biochem Behav 1983;18:47–50.

69 Frohman LA, Muller EE, Cocchi D: Central nervous system mediated inhibition of insulin secretion due to 2-deoxyglucose. Horm Metab Res 1973;5:21–26.

70 Yamamoto H, Katsuya N, Nakagawa H: Role of catecholamine in time-dependent hyperglycemia due to 2-deoxyglucose, mannitol, and glucose. Biomed Res 1983;4: 505–514.

71 Ritter RC, Roelke M, Neville M: Glucoprivic feeding behavior in absence of other signs of glucoprivation. Am J Physiol 1978;255:E617–E621.

72 Bellin SI, Ritter S: Disparate effects of infused nutrients on delayed glucoprivic feeding and hypothalamic norepinephrine turnover. J Neurosci 1981;1:1347–1353.

73 Ritter RC, Slusser PG, Stone S: Glucoreceptors controlling feeding and blood glucose: Location in the hindbrain. Science 1981;213:451–453.

74 Louis-Sylvestre J, LeMagnen J: A fall in blood glucose level precedes meal onset in free-feeding rats. Neurosci Biobehav Rev 1980;4:13–15.

75 Müller K, Scharrer E, Zucker H: Aufhebung eines durch die Mahlzeitenfolge gesetzten 12-Std-Rhythmus der Blutglucosekonzentration durch Änderung der Nahrungszusammensetzung. Naturwissenschaften 1967;54:201–202.

76 Campfield LA, Smith FJ, Brandon P: On-line continuous measurement of blood glucose and meal pattern in free-feeding rats: The role of glucose in meal initiation. Brain Res Bull 1985;14:606–616.

77 Nicolaidis S: Lateral hypothalamic control of metabolic factors related to feeding. Diabetologia 1981;20:426–441.

78 Pollak CP, Green J, Smith GP: Blood glucose prior to meal request in humans isolated from all temporal cues. Physiol Behav 1989;46:529–534.

79 Simon C, Follenius M, Brandenberger G: Postprandial oscillations of plasma glucose, insulin and C-peptide in man. Diabetologia 1987;30:769–773.

80 Ferranini E, Bjorkman O, Reichard GA, et al: The disposal of an oral glucose load in healthy subjects. Diabetes 1985;34:580–588.

81 Tse TF, Clutter WE, Shah SD, et al: Mechanisms of postprandial glucose counterregulation in man. J Clin Invest 1983;72:278–286.

82 Leathwood P, Pollet P: Effects of slow release carbohydrates in the form of bean flakes on the evolution of hunger and satiety in man. Appetite 1988;10:1–11.

83 Flatt JP, Ravussin E, Acheson KJ, et al: Effects of dietary fat on postprandial substrate oxidation and on carbohydrate and fat balances. J Clin Invest 1985;76:1019–1024.

84 Strubbe JH, Steffens AB: Blood glucose levels in portal and peripheral circulation and their relation to food intake in the rat. Physiol Behav 1977;19:303–307.

85 Russek M: A hypothesis on the participation of hepatic glucoreceptors in the control of food intake. Nature 1963;197:79–80.

86 Novin D, Sanderson JO, Van der Weele DA: The effect of isotonic glucose on eating as a function of feeding condition and infusion site. Physiol Behav 1974;13:3–7.

87 Geary N, Scharrer E, Freudlsperger R, et al: Adaptation to high fat diet and carbohydrate-induced satiey in the rat. Am J Physiol 1979;237:R139–R146.

88 Yin TH, Tsai WH, Barone FC, et al: Effects of continuous intramesenteric infusion of glucose and amino acids on food intake of rats. Physiol Behav 1979;22:1207–1210.

89 Woods SC, Stein LJ, McKay LD, et al: Suppression of food intake by intravenous nutrients and insulin in the baboon. Am J Physiol 1984;247:R393–R401.

90 Novin D, Robinson K, Culbreth LA, et al: Is there a role for the liver in control of food intake? Am J Clin Nutr 1985;42:1050–1062.

91 Tordoff MG, Friedman MI: Hepatic portal glucose infusions decrease food intake and increase food preference. Am J Physiol 1986;251:R192–R196.

92 Adair ER, Miller NE, Booth DA: Effects of continuous intravenous infusion of nutritive substances on consummatory behavior in the rat. Commun Behav Biol 1968;A2:25–37.

93 Scharrer E, Thomas DW, Mayer J: Absence of effect of intraaortal glucose infusion upon spontaneous meals of rats. Pflügers Arch 1974;351:315–322.

94 Booth DA, Jarman SP: Inhibition of food intake in the rat following complete absorption of glucose delivered into the stomach, intestine or liver. J Physiol 1976;259:501–522.

95 Walls EK, Koopmans HS: Effect of intravenous nutrient infusions on food intake in rats. Physiol Behav 1989;45:1223–1226.

96 Martyn PA, Hansen BC, Jan KC: The effects of parenteral nutrition on food intake and gastric motility. Nutr Res 1984;33:336–342.

97 Even P, Nicolaidis S: Short-term control of feeding, limitation of the glucostatic theory. Brain Res Bull 1986;17:621–626.

98 Boden G, Shimoyama R, Savage P Jr, et al: Carbohydrate oxidation in patients with type B insulin resistance. Diabetes 1985;34:498–503.

99 Abbot WGH, Howard BV, Christin L, et al: Short-term energy balance: relationship with protein, carbohydrate, and fat balances. Am J Physiol 1988;255:E332–E337.

100 LeMagnen J, Devos M: Metabolic correlates of the meal onset in the free food intake of rats. Physiol Behav 1970;5:805–814.

101 Booth DA, Campbell CS: Relation of fatty acids to feeding behaviour: effects of lighting variation and pent-4-enoate, insulin or propranolol injection. Physiol Behav 1975;15:523–535.

102 Vandermeerschen-Doizé F, Paquay R: Effects of continous long-term intravenous infusion of long-chain fatty acids on feeding behavior and blood components of adult sheep. Appetite 1984;5:137–146.

103 Carpenter RG, Grossman SP: Plasma fat metabolites and hunger. Physiol Behav 1983;30:57–63.

104 Walker DW, Remley NR: The relationships among percentage body weight loss, circulating free fatty acids an consummatory behavior in rats. Physiol Behav 1970;5:301–309.

105 Houpt KA, Reimers TJ, Boyd RD: Changes in free fatty acids and triiodothyronine in response to feeding in pigs. Physiol Behav 1986;37:573–576.

106 Hamilton CL: Problems of refeeding after starvation in the rat. Ann NY Acad Sci 1969;157:1004–1017.

107 Scharrer E, Langhans W: Control of food intake by fatty acid oxidation. Am J Physiol 1986;250:R1003–R1006.

108 LeMagnen J: Body energy balance and food intake: A neuroendocrine regulatory mechanism. Physiol Rev 1983;63:314–386.

109 Bauché F, Sabourault D, Giudicelli Y, et al: 2-Mercaptoacetate administration depresses the β-oxidation pathway through an inhibition of long-chain acyl-CoA dehydrogenase activity. Biochem J 1981;196:803–809.

110 Bauché F, Sabourault D, Giudicelli Y, et al: Inhibition in vitro of acyl-CoA dehydrogenases by 2-mercaptoacetate in rat liver mitochondria. Biochem J 1983;215: 457–464.

111 Friedman MI, Ramirez I, Bowden CR, Tordoff MG: Fuel partitioning and food intake: Role for mitochondrial fatty acid transport. Am J Physiol 1990;258:R216–R221.

112 Kanarek RB, Hirsch JB: Dietary-induced overeating in experimental animals. Fed Proc 1977;36:154–158.

113 Lissner L, Levitsky DA, Strupp BJ, et al: Dietary fat and the regulation of energy intake in human subjects. Am J Clin Nutr 1987;46:886–892.

114 Tremblay A, Plourde G, Despres J-P, Bouchard C: Impact of dietary fat content and fat oxidation on energy intake in humans. Am J Clin Nutr 1989;49:799–805.

115 West DB, Prinz WA, Greenwood MRC: Regional changes in adipose tissue blood flow and metabolism in rats after a meal. Am J Physiol 1989;257:R711–R716.

116 Bremer J, Osmundsen H: Fatty acid oxidation and its regulation; in Numa S (ed): Fatty Acid Oxidation and Its Regulation. Amsterdam, Elsevier, 1984, pp 113–154.

117 Rumpler WV, Seale JL, Miles CW, et al: Energy-intake restriction and diet-composition effects on energy expenditure in men. Am J Clin Nutr 1991;53:430–436.

118 Foltin RW, Fischman MW, Moran TH, et al: Caloric compensation for lunches varying in fat and carbohydrate content by humans in a residential laboratory. Am J Clin Nutr 1990;52:969–980.

119 Metges C, Wolfram G: Medium- and long-chain triglycerides labeled with [13]C: A comparison of oxidation after oral or parenteral administration in humans. J Nutr 1991;121:31–36.

120 Friedman MI, Tordoff MG: Fatty acid oxidation and glucose utilization interact to control food intake in rats. Am J Physiol 1986;251:R840–R845.

121 Even P, Coulaud H, Nicolaidis S: Integrated metabolic control of food intake after 2-deoxy-D-glucose and nicotinic acid injection. Am J Physiol 1988;255:R82–R89.

122 Booth DA: Postabsorptively induced suppression of appetite and the energostatic control of feeding. Physiol Behav 1972;9:199–202.

123 Friedman MI, Stricker ME: The physiological psychology of hunger, a physiological perspective. Psychol Rev 1976;83:409–431.

124 Langhans W, Damaske U, Scharrer E: Different metabolites might reduce food intake by the mitochondrial generation of reducing equivalents. Appetite 1985;6: 143–152.

125 Langhans W, Wiesenreiter F, Scharrer E: Increases in plasma glycerol levels precede

the hypophagia following subcutaneous glycerol injection in rats. Physiol Behav 1983;30:421–424.

126 Langhans W, Wiesenreiter F, Scharrer E: Different effects of subcutaneous *D,L*-3-hydroxybutyrate and acetoacetate injections on food intake in rats. Physiol Behav 486;31:483–486.

127 Langhans W, Pantel K, Scharrer E: Ketone kinetics and *D*-(–)-3-hydroxybutyrate-induced inhibition of feeding in rats. Physiol Behav 1985;34:579–582.

128 Langhans W, Damaske U, Scharrer E: Subcutaneous glycerol injection fails to reduce food intake in rats fed a high protein diet. Physiol Behav 1984;32:785–789.

129 Mauron J, Mottu F, Spohr G: Reciprocal induction and repression of serine dehydratase and phosphoglycerate dehydrogenase by proteins and dietary essential amino acids in rat liver. Eur J Biochem 1973;32:334–342.

130 Begum N, Tepperman HM, Tepperman J: Effects of high fat and high carbohydrate diets on liver pyruvate dehydrogenase and its activation by a chemical mediator released from insulin-treated liver particulate fraction. Effect of neuraminidase treatment on the chemical mediator activity. Endocrinology 1983;112:50–59.

131 Racotta R, Russek M: Food and water intake of rats after intraperitoneal and subcutaneous administration of glucose, glycerol and sodium lactate. Physiol Behav 1977:18:267–273.

132 Rezek M, Novin D: Hepatic-portal nutrient infusion: Effect on feeding in intact and vagotomized rabbits. Am J Physiol 1977;232:E119–E130.

133 Wirtshafter D, Davis JD: Body weight: Reduction by long-term glycerol treatment. Science 1977;198:1271–1273.

134 Glick Z: Food intake of rats administered with glycerol. Physiol Behav 1980;25:621–626.

135 Ramirez I, Friedman M: Glycerol is not a physiologic signal in the control of food intake in rats. Physiol Behav 1982;29:921–925.

136 LeSauter J, Geary N: Pancreatic glucagon and cholecystokinin synergistically inhibit sham-feeding in rats. Am J Physiol 1987;253:R719–R725.

137 Stricker EM, Rowland N, Saller CF, et al: Homeostasis during hypoglycemia: central control of adrenal secretion and peripheral control of feeding. Science 1977;196:79–81.

138 Langhans W, Egli G, Scharrer E: Selective hepatic vagotomy eliminates the hypophagic effect of different metabolites. J Auton Nerv Syst 1985;13:255–262.

139 Rowland N, Bellush LL, Carlton J: Metabolic and neurochemical correlates of glucoprivic feeding. Brain Res Bull 1985;14:617–624.

140 Arase K, Fisler JS, Shargill NS, et al: Intraventricular infusions of 3-hydroxybutyrate and insulin in a rat model of dietary obesity. Am J Physiol 1988;255:R974–R981.

141 Baile CA, Zinn WM, Mayer J: Effects of lactate and other metabolites on food intake of monkeys. Am J Physiol 1970;219:1606–1613.

142 Niewoehner CB, Gilboe DP, Nuttal FQ: Metabolic effects of oral glucose in the liver of fasted rats. Am J Physiol 1984;246:E89–E94.

143 Kimura RE, LaPine TR, Manford Gooch W III: Portal venous and aortic glucose and lactate changes in a chronically catheterized rat. Pediatr Res 1988;23:235–240.

144 Smadja C, Morin J, Ferré P, Girard J: Metabolic fate of a gastric glucose load in unrestrained rats bearing a portal vein catheter. Am J Physiol 1988;254:E407–E413.

145   Langhans W: Hepatic and intestinal handling of metabolites during feeding in rats. Physiol Behav 1991;49:1203–1209.

146   Felig P, Wahren J, Hendler R: Influence of oral glucose ingestion on splanchnic glucose and gluconeogenic substrate metabolism in man. Diabetes 1975;24:468–475.

147   Lovejoy J, Mellen B, DiGirolamo M: Lactate generation following glucose ingestion: Relation to obesity, carbohydrate tolerance and insulin sensitivity. Int J Obes 1990;14:843–855.

148   Newby FD, Wilson LK, Thacker SV, DiGirolamo M: Adipocyte lactate production remains elevated during refeeding after fasting. Am J Physiol 1990;259:E865–E871.

149   Thacker SV, Nickel M, DiGirolamo M: Effects of food restriction on lactate production from glucose by rat adipocytes. Am J Physiol 1987;253:E336–E342.

150   Smyth CJ, Lasichak AG, Levey S: The effect of orally and intravenously administered amino acid mixtures on voluntary food consumption in normal men. J Clin Invest 1947;26:439–445.

151   Mellinkoff S, Frankland S, Boyle D, et al: Relationship between serum amino acid concentration and fluctuations in appetite. J Appl Physiol 1956;8:535–538.

152   Simson PC, Booth DA: Subcutaneous release of amino acid loads on food and water intakes in the rat. Physiol Behav 1973;11:329–336.

153   Anderson GH: Control of protein and energy intake: Role of plasma amino acids and brain neurotransmitters. Can J Physiol Pharmacol 1979;57:1043–1057.

154   Fernstrom JD, Wurtman RJ, Hammerstrom-Wiklund B, et al: Diurnal variations in plasma concentrations of tryptophan, tyrosine and other neutral amino acids: Effect of dietary protein intake. Am J Clin Nutr 1979;32:1912–1922.

155   Peters JC, Harper AE: Protein and energy consumption, plasma amino acid ratios, and brain neurotransmitter concentrations. Physiol Behav 1981;27:287–298.

156   Booth DA, Gibson EL, Baker BJ: Gastromotor mechanism of fenfluramine anorexia. Appetite 1986;7(suppl):57–69.

157   Langhans W, Egli G, Scharrer E: Regulation of food intake by hepatic oxidative metabolism. Brain Res Bull 1985;15:425–428.

158   Sawchenko PE, Friedman MI: Sensory functions of the liver – a review. Am J Physiol 1979;236:R5-R20.

159   Lautt WW: Hepatic nerves: A review of their functions and effects. Can J Physiol Pharmacol 1980;58:105–123.

160   Niijima A: Afferent impulse discharges from glucoreceptors in the liver of the guinea pig. Ann NY Acad Sci 1969;157:690–700.

161   Niijima A: Glucose-sensitive afferent nerve fibres in the hepatic branch of the vagus nerve in the guinea-pig. J Physiol 1982;332:315–323.

162   Niijima A: glucose-sensitive afferent nerve fibers in the liver and their role in food intake and blood glucose regulation. J Auton Nerv Syst 1983;9:207–220.

163   Russek M: Demonstration of the influence of an hepatic glucose-sensitive mechanism on food intake. Physiol Behav 1970;5:1207–1209.

164   Campbell CS, Davis JD: Licking rate in rats is reduced by intraduodenal and intraportal glucose glucose infusion. Physiol Behav 1974;12:357–365.

165   Bellinger LL, Trietley GJ, Bernardis LL: Failure of portal glucose and adrenaline infusions or liver denervation to affect food intake in dogs. Physiol Behav 1976;16:299–304.

166 Geiselman PJ: Food intake following intraduodenal and hepatic-portal infusion of hexoses in the rabbit: evidence that hexose administration can increase subsequent chow intake. Nutr Behav 1984;2:77–87.

167 Tordoff MG, Friedman MI: Hepatic control of feeding: Effect of glucose, fructose, and mannitol infusion. Am J Physiol 1988;254:R969–R976.

168 Tordoff MG, Tluczek JP, Friedman MI: Effect of hepatic portal glucose concentration on food intake and metabolism. Am J Physiol 1989;257:R1474–R1480.

169 Novin D, O'Farrel L, Acevedo-Cruz A, et al: The metabolic bases for 'paradoxical' and normal feeding. Brain Res Bull 1991;27:435–438.

170 Russel PJD, Mogenson GJ: Drinking and feeding induced by jugular and portal infusions of 2-deoxy-D-glucose. Am J Physiol 1975;229:1014–1018.

171 Tordoff MG, Hopfenbeck J, Novin D: Hepatic vagotomy (partial hepatic denervation) does not alter ingestive responses to metabolic challenges. Physiol Behav 1982; 28:417–424.

172 Delprete E, Scharrer E: Hepatic branch vagotomy attenuates the feeding response to 2-deoxy-D-glucose in rats. Exp Physiol 1990;75:259–261.

173 Campfield LA, Smith FJ: Systemic factors in the control of food intake – Evidence for patterns as signals; in Stricker EM (ed): Handbook of Behavioral Neurobiology. Neurobiology of Food and Fluid Intake. New York, Plenum Press, 1990, vol 10, pp 183–206.

174 Hermann GE, Rogers RC: Convergence of vagal and gustatory afferent input within the parabrachial nucleus of the rat. J Auton Nerv Syst 1985;13:1–17.

175 Grill HJ: Caudal brainstem integration of taste and internal state factors in behavioral and autonomic responses; in Kare MR, Brand JG (eds): Interaction of the Chemical Senses with Nutrition. Orlando, Academic Press, 1986, pp 347–371.

176 Powley TL, Berthoud H-R: Participation of the vagus and other autonomic nerves in control of food intake; in Ritter RC, Ritter S, Barnes CD (eds): Feeding Behavior, Neural and Humoral Control. Orlando, Academic Press, 1986, pp 67–102.

177 Oomura Y, Yoshimatsu H: Neural network of glucose monitoring system. J Auton Nerv Syst 1984;10:359–372.

178 Shimizu N, Oomura Y, Novin D, et al: Functional correlations between lateral hypothalamic glucose-sensitive neurons and hepatic portal glucose-sensitive units in the rat. Brain Res 1983;265:49–54.

179 Schmitt M: Influences of hepatic portal receptors on hypothalamic feeding and satiety centers. Am J Physiol 1973;225:1089–1095.

180 Rogers RC, Kajrilas PJ, Hermann GE: Projection of the hepatic branch of the splanchnic nerve to the brainstem of the rat. J Auton Nerv Syst 1984;11:223–225.

181 Langhans W, Scharrer E: Evidence for a vagally-mediated satiety signal derived from hepatic fatty acid oxidation. J Auton Nerv Syst 1987;18:13–18.

182 Ritter S, Taylor JS: Capsaicin abolishes lipoprivic but not glucoprivic feeding in rats. Am J Physiol 1989;256:R1232–R1239.

183 Ritter S, Taylor JS: Vagal sensory neurons are required for lipoprivic but not glucoprivic feeding in rats. Am J Physiol 1990;258:R1395–R1401.

184 Seifter S, Englard S: Energy metabolism; in Arias IM, Jakoby WB, Popper H, Schachter D, Shafritz DA (eds): The Liver: Biology and Pathobiology. New York, Raven Press, 1988, pp 279–315.

185 Radziuk J, McDonald TJ, Rubenstein D, et al: Initial splanchnic extraction of ingested glucose in normal man. Metabolism 1978;27:657–669.

186 Hems DA, Whitton PO, Taylor EA: Glycogen synthesis in the perfused liver of the starved rat. Biochem J 1972;129:529–538.

187 Newgard CB, Moore SV, Foster DW, et al: Efficient hepatic glycogen synthesis in refeeding rats requires continued carbon flow through the gluconeogenetic pathway. J Biol Chem 1983;259:6948–6963.

188 Radziuk J: Hepatic glycogen in humans. I. Direct formation after oral and intravenous glucose or after a 24-h fast. Am J Physiol 1989;257:E145–E157.

189 Radziuk J: Hepatic glycogen in humans. II. Gluconeogenetic formation after oral and intravenous glucose. Am J Physiol 1989;257:E158–E169.

190 Foster DW: From glycogen to ketones – and back. Diabetes 1984;33:1188–1199.

191 McGarry JD, Kuwajima M, Newgard CB, et al: From dietary glucose to liver glycogen: the full circle round. Annu Rev Nutr 1987;7:51-73.

192 Giesecke D: Species differences relevant to nutrition and metabolism research; in Dietze G, Grünert A, Kleinberger G, Wolfram G (eds): Clinical and Metabolic Research. Basel, Karger, 1986, pp 311–328.

193 Dobson GP, Veech RL, Passonneau JV, et al: In vivo portal-hepatic venous gradients of glycogenic precursors and incorporation of D-(3-^{3}H)glucose into liver glycogen in the awake rat. J Biol Chem 1990;265:16350–16357.

194 Langhans W, Geary N, Scharrer E: Liver glycogen content decreases during meals in rats. Am J Physiol 1982;243:R450–R453.

195 Wolever TMS: Metabolic effects of continuous feeding. Metabolism 1990;39:947–951.

196 Langhans W, Pantel K, Müller-Schell W, et al: Hepatic handling of pancreatic glucagon and glucose during meals in rats. Am J Physiol 1984;247:R827–R832.

197 Davis MA, Williams PE, Cherrington AD: Effect of a mixed meal on hepatic lactate and gluconeogenetic precursor metabolism in dogs. Am J Physiol 1984;247:E362–E369.

198 Van de Werve G, Jeanrenaud B: The onset of liver glycogen synthesis in fasted-refed lean and genetically obese (fa/fa) rats. Diabetologia 1987;30:169–174.

199 Hue L, Sobrino F, Bosca L: Difference in glucose sensitivity of liver glycolysis and glycogen synthesis – relationship between lactate production and fructose 2,6-biphosphate concentration. Biochem J 1984;224:779–786.

200 Hue L, Rider MH: Role of fructose-2,6-biphosphate in the control of glycolysis in mammalian tissues. Biochem J 1987;245:313–324.

201 Claus TH, Nyfeler F, Muenkel HA, et al: Changes in key regulatory enzymes of hepatic carbohydrate metabolism after glucose loading of starved rats. Biochem Biophys Res Commun 1984;155:655–661.

202 Katz N, Jungermann K: Autoregulatory shift from fructolysis to lactate gluconeogenesis in rat hepatocyte suspensions. The problem of metabolic zonation of liver parenchyma. Hoppe Seyler's Z Physiol Chem 1976;357:359–375.

203 Jungermann K: Functional heterogeneity of periportal and perivenous hepatocytes. Enzyme 1986;35:161–180.

204 Rich-Denson C, Kimura RE: Evidence in vivo that most of the intraluminally absorbed glucose is absorbed intact into the portal vein and not metabolized to lactate. Biochem J 1988;254:931–934.

205 Shoemaker WC, Yanof HM, Turk LN III, et al: Glucose and fructose absorption in the unanesthetized dog. Gastroenterology 1963;44:654–663.

206 Fafournoux P, Demigné C, Rémésy C: Carrier-mediated uptake of lactate in rat hepatocytes. J Biol Chem 1985;260:292–299.

207 Holness MJ, Sugden MC: Pyruvate dehydrogenase activities during the fed-to-starved transition and on re-feeding after acute or prolonged starvation. Biochem J 1989;258:529–533.

208 Rognstad R: The role of mitochondrial pyruvate transport in the control of lactate gluconeogenesis. Int J Biochem 1983;15:1417–1421.

209 Agius L, Alberti KGMM: Regulation of flux through pyruvate dehydrogenase and pyruvate carboxylase in rat hepatocytes, effects of fatty acids and glucagon. Eur J Biochem 1985;152:699–707.

210 Randle PJ: Fuel selection in animals. Biochem Soc Trans 1986;14:799–806.

211 Pi-Sunyer FX, Hashim SA, Van Itallie TB: Insulin and ketone responses to ingestion of medium and long-chain triglycerides in man. Diabetes 1969;18:96–100.

212 Scharrer E, Langhans W: Metabolic and hormonal factors controlling food intake. Int J Vitam Nutr Res 1988;58:249–261.

213 Geary N, Langhans W, Scharrer E: Metabolic concomitants of glucagon-induced suppression of feeding in the rat. Am J Physiol 1981;241:R330–R335.

214 Langhans W, Zieger U, Scharrer E, et al: Stimulation of feeding in rats by intraperitoneal injection of antibodies to glucagon. Science 1982;218:894–896.

215 Geary N, Smith GP: Selective hepatic vagotomy blocks pancreatic glucagon's satiety effect. Physiol Behav 1983;31:391–394.

216 Langhans W, Scharrer E: Evidence for a role of the sodium pump of hepatocytes in the control of food intake. J Auton Nerv Syst 1987;20:199–205.

217 Balaban RS: Regulation of oxidative phosphorylation in the mammalian cell. Am J Physiol 1990;258:C377–C389.

218 Coulson RA: Metabolic rate and the flow theory: a study in chemical engineering. Comp Physiol Psychol 1986;84A:217–229.

219 Hernandez LA, Kvietys PR, Granger DN: Postprandial hemodynamics in the conscious rat. Am J Physiol 1986;251:G117–G123.

220 Milligan LP, McBride BW: Energy cost of ion pumping by animal tissues. J Nutr 1985;115:1374–1382.

221 Swaminathan R, Chan ELP, Sin LY, et al: The effect of ouabain on metabolic rate in guinea-pigs: Estimation of energy cost of sodium pump activity. Br J Nutr 1989;61:467–473.

222 Erecinska M, Wilson DF: Regulation of cellular energy metabolism. J Membr Biol 1982;70:1–14.

223 Soboll S, Seitz HJ, Sies H, et al: Effect of long-chain fatty acyl-CoA on mitochondrial and cytosolic ATP/ADP ratios in the intact liver cell. Biochem J 1984;220:371–376.

224 Plomp PJAM, van Roermund CWT, Groen AK, et al: Mechanism of the stimulation of respiration by fatty acids in rat liver. FEBS Lett 1985;193:243–246.

225 Jones DP: Intracellular diffusion gradients of O_2 and ATP. Am J Physiol 1986;250:C663–C675.

226 Tordoff MG, Rafka R, DiNovi MJ, et al: 2,5-Anhydro-D-mannitol: a fructose analogue that increases food intake in rats. Am J Physiol 1988;254:R150–R153.

227 Hanson RL, Ho RS, Wiseberg JJ, et al: Inhibition of gluconeogenesis and glycogenolysis by 2,5-anhydro-D-mannitol. J Biol Chem 1984;259:218–223.

228   Stevens HC, Covey TR, Dills WL Jr: Inhibition of gluconeogenesis by 2,5-anhydro-
      D-mannitol in isolated rat hepatocytes. Biochim Biophys Acta 1985;845:502–506.
229   Fitz JG, Scharschmidt BF: Regulation of transmembrane electrical potential gra-
      dient in rat hepatocytes in situ. Am J Physiol 1987;252:G56–G64.
230   Russek M, Grinstein S: Coding of metabolic information by hepatic glucoreceptors;
      in Myers RD, Drucker-Colin RR (eds): Neurohumoral Coding of Brain Function.
      New York, Plenum Publishing, 1974, pp 81–97.
231   Dambach G, Friedmann N: Substrate-induced membrane potential changes in the
      perfused rat liver. Biochim Biophys Acta 1974;367:366–370.
232   Meyer AH, Scharrer E: Hyperpolarization of the cell membrane of mouse hepato-
      cytes by metabolizable and non-metabolizable monosaccharides. Physiol Behav
      1991;50:351–355.
233   Cohen BJ, Lechene C: Alanine stimulation of passive potassium efflux in hepato-
      cytes is independent of Na^+-K^+ pump activity. Am J Physiol 1990;258:C24–C29.
234   Probst I, Unthan-Fechner K: Activation of glycolysis by insulin with a sequential
      increase of the 6-phosphofructo-2-kinase activity, fructose-2,6-biphosphate level
      and pyruvate kinase activity in cultured rat hepatocytes. Eur J Biochem 1985;153:
      347–353.
235   Terrettaz J, Assimacopoulos-Jeannet F, Jeanrenaud B: Inhibition of hepatic glucose
      production by insulin in vivo in rats: contribution of glycolysis. Am J Physiol 1986;
      250:E346–E351.
236   Tanaka K, Inoue S, Takamura Y, et al: Arginine sensors in the hepato-portal system
      and their reflex effects on pancreatic efferents in the rat. Neurosci Lett 1986;72:
      69–73.
237   Tanaka K, Inoue S, Nagase H, et al: Amino acid sensors sensitive to alanine and
      leucine exist in the hepato-portal system in the rat. J Auton Nerv Syst 1990;31:
      41–46.
238   Niijima A: Reflex control of the autonomic nervous system activity from the glucose
      sensors in the liver in normal and midpontine-transected animals. J Auton Nerv Syst
      1984;10:279–285.
239   Sakaguchi T, Shimojo E: Inhibition of gastric motility induced by hepatic portal
      injections of *D*-glucose and its anomers. J Physiol (Lond) 1984;351:573–581.
240   Adachi A, Niijima A: Thermosensitive afferent fibers in the hepatic branch of the
      vagus nerve in the guinea pig. J Auton Nerv Syst 1982;5:101–109.
241   Adachi A: Thermosensitive and osmoreceptive afferent fibers in the hepatic branch
      of the vagus nerve. J Auton Nerv Syst 1984;10:269–273.
242   Di Bella L, Tarozzi G, Rossi MT, et al: Effect of liver temperature increase on food
      intake. Physiol Behav 1981;26:45–51.
243   Di Bella L, Tarozzi G, Rossi MT, et al: Behavioral patterns proceeding from liver
      thermoreceptors. Physiol Behav 1981;26:53–59.
244   Wilhelmj CM, Bollman JL, Mann FC: Studies on the physiology of the liver. XVII.
      The effect of removal of the liver on the specific dynamic action of amino acids
      administered intravenously. Am J Physiol 1928;87:387–401.
245   Strubbe JH, Alingh Prins AJ: Control of feeding behavior by core temperature in
      rats (abstract). Appetite 1986;7:302.
246. Westerterp-Plantenga LS, Wouters L, Ten Hoor F: Deceleration in cumulative food
      intake curves, changes in body temperature and diet-induced thermogenesis. Physiol
      Behav 1990;48:831–836.

247 Berry MN, Clark DG, Grivell AR, et al: The calorigenic nature of hepatic ketogenesis: an explanation for the stimulation of respiration induced by fatty acid substrates. Eur J Biochem 1983;131:205–214.

248 Berry MN, Clark DG, Grivell AR, et al: The contribution of hepatic metabolism to diet-induced thermogenesis. Metabolism 1985;34:141–147.

249 Rottenberg H, Hashimoto K: Fatty acid uncoupling of oxidative phosphorylation in rat liver mitochondria. Biochemistry 1986;25:1747–1755.

250 Bellinger LL, Williams FE: The effects of liver denervation on food and water intake in the rat. Physiol Behav 1980;26:663–671.

251 Louis-Sylvestre J, Servant J-M, Molimard R, LeMagnen J: Effect of liver denervation on the feeding pattern of rats. Am J Physiol 1980;239:R66–R70

252 Bellinger LL, Mendel VE, Williams FE, et al: The effect of liver denervation on meal patterns, body weight and body composition of rats. Physiol Behav 1984;33:661–667.

253 Friedman MI, Sawchenko PE: Evidence for hepatic involvement in control of ad libitum food intake in rats. Am J Physiol 1984;247:R106–R113.

254 Kamada N, Calne RY: Orthotopic liver transplantation in the rat. Transplantation 1979;28:47–50.

255 Louis-Sylvestre J, Larue-Achagiotis C, Michel A, et al: Feeding pattern of liver-transplanted rats. Physiol Behav 1990;48:321–326.

256 Deutsch JA, Jang Ahn S: The splanchnic nerve and food intake regulation. Behav Neural Biol 1986;45:43–47.

257 Kraly FS, Jerome C, Smith GP: Specific postoperative syndromes after total and selective vagotomies in the rat. Appetite 1986;7:1–17.

258 Kasser TR, Harris RBS, Martin RJ: Level of satiety: fatty acid and glucose metabolism in three brain sites associated with feeding. Am J Physiol 1985;248:R447–R452.

259 Kasser TR, Harris RBS, Martin RJ: Level of satiety: GABA and pentose shunt activities in three brain sites associated with feeding. Am J Physiol 1985;248:R453–R458.

260 Booth DA: Conditioned satiety in the rat. J Comp Physiol Psychol 1972;81:457–471.

261 Deutsch JA, Tabuena JA: Learning of gastrointestinal satiety signals. Behav Neural Biol 1986;45:292–299.

262 Langhans W, Kunz U, Scharrer E: Hepatic vagotomy increases consumption of a novel-tasting diet. Physiol Behav 1989;46:671–678.

263 Tordoff MG, Ulrich PM, Sandler F: Flavor preferences and fructose: Evidence that the liver detects the unconditioned stimulus for calorie-based learning. Appetite 1990;14:29–44.

264 Kunz-Straumann U: Untersuchungen zum Einfluss der hepatischen Vagotomie auf die Regulation der Nahrungsaufnahme. Vet Med Thesis, University of Zürich, 1987.

265 Sjoström L, Garellick G, Krotkiewski A, et al: Peripheral insulin in response to the sight and smell of food. Metabolism 1980;29:901–909.

266 Sahakian BJ, Lean ME, Robbins TW, et al: Salivation and insulin secretion in response to food in non-obese men and women. Appetite 1981;2:209–216.

267 Bellisle F, Louis-Sylvestre J, Demozay F, et al: Reflex insulin response associated to food intake in human subjects. Physiol Behav 1983;31:515–521.

268 Strubbe JH, Steffens AB: Rapid insulin release after ingestion of a meal in the un-anesthetized rat. Am J Physiol 1975;229:1019–1022.

269 Louis-Sylvestre J: Preabsorptive insulin release and hypoglycemia in rats. Am J Physiol 1976;230:56–60.

270 De Jong A, Strubbe JH, Steffens AB: Hypothalamic influence on insulin and gluca-gon release in the rat. Am J Physiol 1977;233:E380–E388.

271 Louis-Sylvestre J, LeMagnen J: Palatability and preabsorptive insulin release. Neu-rosci Behav Rev 1980;4(suppl 1):43–46.

272 Woods SC, Vaselli JR, Kaestner E, et al: Conditioned insulin secretion and meal feeding in rats. J Comp Physiol Psychol 1977;91:128–133.

273 Creutzfeldt W: The incretin concept today. Diabetologia 1979;16:75–85.

274 Day JL, Johansen K, Ganda OP, et al: Factors governing insulin and glucagon responses during normal meals. Clin Endocrinol 1978;9:443–454.

275 Berthoud HR: The relative contribution of the nervous system, hormones, and metabolites to the total insulin response during a meal in the rat. Metabolism 1984;33:18–25.

276 Grossman MI, Stein IF: Vagotomy and the hunger-producing action of insulin in man. J Appl Physiol 1948;1:263–267.

277 MacKay EM, Callaway JW, Barnes RH: Hyperalimentation in normal animals pro-duced by protamine zinc insulin. Nutrition 1940;20:59–66.

278 Brandes JS: Insulin-induced overeating in the rat. Physiol Behav 1977;18:1095–1102.

279 Ritter RC, Balch OK: Feeding in response to insulin but not 2-deoxy-glucose in the hamster. Am J Physiol 1978;234:E20–E24.

280 Rowland N: Glucoregulatory feeding in cats. Physiol Behav 1981;26:901–903.

281 Steffens AB: The influence of insulin injections and infusions on eating and blood glucose levels in the rat. Physiol Behav 1969;4:823–828.

282 Flynn FW, Grill HJ: Insulin elicits ingestion in decerebrate rats. Science 1983;221:188–190.

283 Hyde TM, Miselis RR: Effect of area postrema/caudal medial nucleus of the solitary tract lesions on food intake and body weight. Am J Physiol 1983;244:R577–R587.

284 Oldendorf WH: Brain uptake of radiolabeled amino acids, amines, and hexoses after arterial injection. Am J Physiol 1971;221:1629–1639.

285 Rowland N, Stricker EM: Differential effects of glucose and fructose infusions on insulin-induced feeding in rats. Physiol Behav 1979;22:387–389.

286 Rodin J, Wack J, Ferrannini E, et al: Effect of insulin and glucose on feeding behav-ior. Metabolism 1985;34:826–831.

287 Kott JN, Kenney NJ, Bhatia AJ, et al: Response to chronic insulin administration: Effect of area postrema ablation. Physiol Behav 1989;46:971–976.

288 Steffens AB: Plasma insulin content in relation to blood glucose level and meal pattern in the normal and hypothalamic hyperphagic rat. Physiol Behav 1969;5:147–151.

289 Cahill GF: Obesity and insulin levels. N Engl J Med 1971;284:1268–1269.

290 Stern JS, Batchelor BR, Hollander N, et al: Adipose cell size and immunoreactive insulin level in obese and normal weight adults. Lancet 1972;ii:948–951.

291 Bray GA, York DA: Hypothalamic and genetic obesity in experimental animals: an autonomic and endocrine hypothesis. Physiol Rev 1979;59:719–809.

292 Jeanrenaud B: Insulin and obesity. Diabetologia 1979;17:133–138.

293 Jeanrenaud B: A hypothesis on the aetiology of obesity: Dysfunction of the central nervous system as a primary cause. Diabetologia 1985;28:502–513.

294 Hansen FM, Nilsson P, Hustvedt BE, et al: Significance of hyperinsulinemia in ventromedial hypothalamus-lesioned rats. Am J Physiol 1983;244:E203–E208.

295 King BM, Frohman LA: The role of vagally-mediated hyperinsulinemia in hypothalamic obesity. Neurosci Biobehav Rev 1982;6:205–214.

296 Larue-Achagiotis C, LeMagnen J: Effect of long-term insulin on body weight and food intake: Intravenous versus intraperitoneal routes. Appetite 1985;6:319–329.

297 Geary N, Grötschel H, Petry H, et al: Meal patterns and bodyweight changes during insulin hyperphagia and postinsulin hypophagia. Behav Neural Biol 1981;31:435–442.

298 Booth DA: Some characteristics of feeding during streptozotocin-induced diabetes in the rat. J Comp Physiol Psychol 1972;80:238–249.

299 Thomas DW, Scharrer E, Mayer J: Effects of alloxan-induced diabetes on the feeding patterns of rats. Physiol Behav 1976;17:345–349.

300 Friedman MI: Hyperphagia in rats with experimental diabetes melitus: Response to a decreased supply of utilizable fuels. J Comp Physiol Psychol 1978;92:109–117.

301 Van der Weele DA, Pi-Sunyer FX, Novin D, et al: Chronic insulin infusion suppresses food ingestion and body weight gain in rats. Brain Res Bull 1980;5(suppl 4):7–11.

302 Van der Weele DA, Harackiewicz E, Van Itallie TB: Elevated insulin and satiety in obese and normal-weight rats. Appetite 1982;3:99–109.

303 Anika SM, Houpt TL, Houpt KA: Insulin as a satiety hormone. Physiol Behav 1980;25:21–23.

304 Deetz LE, Wangsness PJ, Kavanaugh JF, et al: Effect of intraportal and continuous intrajugular administration of insulin on feeding in sheep. J Nutr 1980;110:1983–1991.

305 Deetz LE, Wangsness PJ: Influence of intrajugular administration of insulin, glucagon and propionate on voluntary feed intake of sheep. J Anim Sci 1981;53:427–433.

306 Van der Weele DA, Harackiewicz E, Vaselli JR: Tolbutamide affects food ingestion in a manner consistent with its glycemic effects in the rat. Physiol Behav 1988;44:679–683.

307 Lindberg NO, Coburn C, Stricker EM: Increased feeding by rats after subdiabetogenic streptozotocin treatment: A role for insulin in satiety. Behav Neurosci 1984;98:138–145.

308 Oetting RL, Van der Weele DA: Insulin suppresses intake without inducing illness in sham-feeding rats. Physiol Behav 1985;34:557–562.

309 Posner BI: Insulin interaction with the central nervous system: Nature and possible significance. Proc Nutr Soc 1987;46:97–103.

310 Duffy KR, Pardridge WM: Blood-brain barrier transcytosis of insulin in developing rabbits. Brain Res 1987;420:32–38.

311 Woods SC, Lotter ED, McKay LD, et al: Chronic intracerebroventricular infusion of insulin reduces food intake and body weight of baboons. Nature 1979;282:503–505.

312 Brief DJ, Davis JD: Reduction of food intake and body weight by chronic intraventricular insulin infusion. Brain Res Bull 1984;12:571–575.

313 Plata-Salaman CR, Oomura Y, Shimizu N: Dependence of food intake on acute and chronic ventricular administration of insulin. Physiol Behav 1986;37:717–734.

314 McGowan MK, Andrews KM, Kelly J, Grossman SP: Effects of chronic intrahypothalamic infusion of insulin on food intake and diurnal meal patterning in the rat. Behav Neurosci 1990;104:371–383.

315 Strubbe JH, Mein CG: Increased feeding in response to bilateral injection of insulin antibodies in the VMH. Physiol Behav 1977;19:309–313.

316 Giza BK, Scott TR: Blood glucose level affects perceived sweetness intensity in rats. Physiol Behav 1987;41:459–464.

317 Giza BK, Scott TR: Intravenous insulin infusions in rats decrease gustatory-evoked responses to sugars. Am J Physiol 1987;252:R994–R1002.

318 Lovett D, Booth DA: Four effects of insulin on food intake. Q J Exp Psychol 1970; 22:406–419.

319 Cherrington AD, Williams PE, Abou-Mourad N, et al: Insulin as a mediator of hepatic glucose uptake in the conscious dog. Am J Physiol 1982;242:E97–E101.

320 Christ B, Probst I, Jungermann K: Antagonistic regulation of the glucose/glucose 6-phosphate cycle by insulin and glucagon in cultured hepatocytes. Biochem J 1986; 238:185–191.

321 Bessman SP, Mohan C, Zaidise I: Intracellular site of insulin action: mitochondrial Krebs cycle. Proc Natl Acad Sci USA 1986;83:5067–5070.

322 Geary N: Pancreatic glucagon signals postprandial satiety. Neurosci Biobehav Rev 1990;14:323–328.

323 Ishida T, Chou J, Lewis RM, et al: The effect of ingestion of meat on hepatic extraction of insulin and glucagon and hepatic glucose output in conscious dogs. Metabolism 1983;32:558–567.

324 Denker H, Hedner P, Holst J, et al: Pancreatic glucagon response to an ordinary meal. Scand J Gastroenterol 1975;10:471–474.

325 Bloom SR, Edwards AV: The release of pancreatic glucagon and inhibition of insulin in response to stimulation of the sympathetic innervation. J Physiol 1975;253:157–173.

326 LeSauter J, Noh U, Geary N: Hepatic portal infusion of glucagon antibodies increases spontaneous meal size in rats. Am J Physiol 1991;261:R162–R165.

327 Stunkard AJ, Van Itallie TB, Reis BB: The mechanism of satiety: Effects of glucagon on gastric hunger contractions in man. Proc Soc Exp Biol Med 1955;89:258–261.

328 Penick SB, Hinkle LE: Depression of food intake induced in healthy subjects by glucagon. N Engl J Med 1961;264:893–897.

329 Davidson IWF, Salter JM, Best CH: The effect of glucagon on the metabolic rate of rats. Am J Clin Nutr 1960;8:540–546.

330 Holloway SA, Stevenson JAF: Effect of glucagon on food intake and weight gain in the young rat. Can J Physiol 1964;42:867–872.

331 Martin JR, Novin D: Decreased feeding in rats following hepatic-portal infusion of glucagon. Physiol Behav 1977;19:461–466.

332 LeSauter J, Geary N: Hepatic portal glucagon infusion decreases spontaneous meal size in rats. Am J Physiol 1991;261:R154–R161.

333 Langhans W, Scharrer E, Geary N: Pancreatic glucagon's effects on satiety and hepatic glucose release are independently affected by diet composition. Physiol Behav 1986;36:483–487.

334 Martin JR, Novin D, Van der Weele DA: Loss of glucagon suppression of feeding after vagotomy in rats. Am J Physiol 1978;234:E314–E318.

335 Van der Weele DA, Geiselman PJ, Novin D: Pancreatic glucagon, food deprivation and feeding in intact and vagotomized rabbits. Physiol Behav 1979;23:155–158.

336 MacIsaac L, Geary N: Partial liver denervations dissociate the inhibitory effects of pancreatic glucagon and epinephrine on feeding. Physiol Behav 1985;35:233–237.

337 Weatherford SC, Ritter S: Glucagon satiety: Diurnal variation after hepatic branch vagotomy or intraportal alloxan. Brain Res Bull 1986;17:545–549.

338 Bellinger LL, Williams FE: Glucagon and epinephrine suppression of food intake in liver denervated rats. Am J Physiol 1986;251:R349–R358.

339 Ritter S, Weatherford SC: Capsaicin pretreatment blocks glucagon-induced suppression of food intake. Appetite 1986;7:291.

340 Weatherford SC, Ritter S: Lesion of vagal afferent terminals impairs glucagon-induced suppression of food intake. Physiol Behav 1988;43:645–650.

341 Van der Weele DA, Harackiewicz E, Di Conti M: Pancreatic glucagon administration, feeding, glycemia, and liver glycogen in rats. Brain Res Bull 1980;5(suppl 4): 17–21.

342 Ritter S, Weatherford SC, Stone SL: Glucagon-induced inhibition of feeding is impaired by hepatic portal alloxan injection. Am J Physiol 1986;250:R682–R690.

343 Geary N, Smith GP: Pancreatic glucagon fails to inhibit sham feeding in the rat. Peptides 1982;3:163–166.

344 Langhans W, Duss M, Scharrer E: Decreased feeding and supraphysiological plasma levels of glucagon after glucagon injection in rats. Physiol Behav 1987;41:31–35.

345 Geary N, Farhoody N, Gersony A: Food deprivation dissociates pancreatic glucagon's effects on satiety and hepatic glucose production at dark onset. Physiol Behav 1987;39:507–511.

346 Hers H, Hue L: Gluconeogenesis and related aspects of glycolysis. Annu Rev Biochem 1983;52:617–653.

347 Davis MA, Williams PE, Cherrington AD: Effect of glucagon on hepatic lactate metabolism in the conscious dog. Am J Physiol 1985;248:E463–E470.

348 Van de Werve G: Liver glucose-6-phosphatase activity is modulated by physiological intracellular Ca^{2+} concentrations. J Biol Chem 1989;264:6033–6036.

349 Weick BG, Ritter S: Stimulation of insulin release and suppression of feeding by hepatic portal glucagon infusion in rats. Physiol Behav 1986;38:531–536.

350 Yamazaki RK: Glucagon stimulation of mitochondrial respiration. J Biol Chem 1975;250:7924–7930.

351 McGarry JD, Foster DW: Glucagon and ketogenesis; in Lefebvre PJ (ed): Glucagon I. Berlin, Springer, 1983, pp 383–398.

352 Petersen OH: The effect of glucagon on the liver cell membrane potential. J Physiol 1974;239:647–656.

353 Friedmann N, Dambach G: Antagonistic effects of insulin on glucagon-evoked hyperpolarization, a correlation between changes in membrane potential and gluconeogenesis. Biochim Biophys Acta 1980;596:180–185.

354 Edmondson JW, Miller BA, Lumeng L: Effect of glucagon on hepatic taurocholate uptake: Relationship to membrane potential. Am J Physiol 1985;249:G427–G433.

355 Kimura S, Suzaki T, Kobayashi S, et al: Effects of glucagon on the redox states of cytochromes in mitochondria in situ in perfused rat liver. Biochem Biophys Res Commun 1984;119:212–219.

356 McCormack JG, Assimacopoulos-Jeannet FD, Denton RM: The effect of Ca-mobilizing hormones on the inner-mitochondrial Ca^{2+}-sensitive dehydrogenases in the liver; in Kraus-Friedmann N (ed): Hormonal Control of Gluconeogenesis. Boca Raton, CRC Press, 1986, pp 81–98.

357 Pecker F, Lottersztajn S, Epand R, et al: Hormonal effects on the hepatic plasma membrane Ca^{2+} pump; in Kraus-Friedmann N (ed): Hormonal Control of Gluconeogenesis. Boca Raton, CRC Press, 1986, pp 89–103.

358 Mallat A, Pavoine C, Dugfour M, et al: A glucagon fragment is responsible for the inhibition of the liver Ca^{2+} pump by glucagon. Nature 1987;325:620–622.

359 Geary N: Glucagon-(1-21) fails to inhibit meal size in rats. Peptides 1987;8:943–945.

360 Kraus-Friedmann N, Hummel L, Radominska-Pyrek A, et al: Glucagon stimulation of Na$^+$,K$^+$-ATPase. Mol Cell Biochem 1982;44:173–180.

361 Oomura Y, Ono T, Ooyama H, et al: Glucose and osmosensitive neurones of the rat hypothalamus. Nature 1969;222:282–284.

362 Oomura Y, Ooyama H, Sugimori M, et al: Glucose inhibition of glucose-sensitive neurones in the rat lateral hypothalamus. Nature 1974;247:284–286.

363 Mizuno Y, Oomura Y: Glucose responding neurons in the nucleus tractus solitarius of the rat: In vitro study. Brain Res 1984;307:109–116.

364 Adachi A, Kobashi M, Miyoshi N, et al: Chemosensitive neurons in the area postrema of the rat and their possible functions. Brain Res Bull 1991;26:137–140.

365 Miselis RR, Epstein AN: Feeding induced by intracerebroventricular 2-deoxy-D-glucose in the rat. Am J Physiol 1975;229:1438–1447.

366 Berthoud HR, Mogenson GJ: Ingestive behavior after intracerebral and intracerebroventricular infusions of glucose and 2-deoxy-D-glucose. Am J Physiol 1977;233: R127–R133.

367 Ritter RC, Slusser PG: Feeding and hyperglycemia induced by 5-thioglucose. Am J Physiol 1980;238:E141–E144.

368 Murnane JM, Ritter S: Intraventricular alloxan impairs feeding to both central and systemic glucoprivation. Physiol Behav 1985;34:609–613.

369 Grossman SP: The role of glucose, insulin and glucagon in the regulation of food intake and body weight. Neurosci Behav Rev 1986;10:295–315.

370 Davis JD, Wirtshafter D, Asin KE, et al: Sustained intracerebroventricular infusion of brain fuels reduces body weight and food intake in rats. Science 1981;212:81–82.

371 Kurata K, Fujimoto K, Sakata T, et al: D-Glucose suppression of eating after intra-third ventricle infusion in rat. Physiol Behav 1986;37:615–620.

372 Tsujii S, Bray GA: Effects of glucose, 2-deoxyglucose, phlorizin, and insulin on food intake of lean and fatty rats. Am J Physiol 1990;258:E476–E481.

373 Oomura Y: Input-output organization in the hypothalamus relating to food intake behavior; in Morgane PJ, Panksepp J (eds): Handbook of the Hypothalamus. Physiology of the Hypothalamus. New York, Dekker, 1980, vol 2, pp 557–620.

374 Stellar E: The physiology of motivation. Psychol Rev 1954;61:5–22.

375 Sakaguchi T, Bray GA: The effect of intrahypothalamic injections of glucose on sympathetic efferent firing rate. Brain Res Bull 1987;18:591–595.

376 Smythe GA, Grunstein HS, Bradshaw JE, et al: Relationships between brain noradrenergic activity and blood glucose. Nature 1984;308:65–67.

377 Chafetz MD, Parko K, Diaz S, et al: Relationships between medial hypothalamic α_2-receptor binding, norepinephrine, and circulating glucose. Brain Res 1986;384. 404–408.

378 Jhanwar-Uniyal M, Papamichael MJ, Leibowitz SF: Glucose-dependent changes in α_2-noradrenergic receptors in hypothalamic nuclei. Physiol Behav 1988;44:611–617.

379 Leibowitz SF: Hypothalamic paraventricular nucleus: interaction between α_2-noradrenergic system and circulating hormones and nutrients in relation to energy balance. Neurosci Biobehav Rev 1988;12:101–104.

380 Nobegra JN, Coscina DV: Regional changes in ^{14}C-2-deoxyglucose uptake after feeding-inducing intrahypothalamic norepinephrine injections. Brain Res Bull 1990;24:249–255.

381 Himmi T, Boyer A, Orsini JC: Changes in lateral hypothalamic neuronal activity accompanying hyper- and hypoglycemias. Physiol Behav 1988;44:347–354.

382 Orsini JC, Himmi T, Wiser AK, et al: Local versus indirect action of glucose on the lateral hypothalamic neurons sensitive to glycemic level. Brain Res Bull 1990;25: 49–53.

383 Shimizu H, Bray GA: Hypothalamic monoamines measured by mirodialysis in rats treated with 2-deoxy-glucose or d-fenfluramine. Physiol Behav 1989;46:799–807.

384 Kimura H, Kuriyama K: Distribution of gamma-aminobutyric acid in the rat hypothalamus: Functional correlates with activities of appetite-controlling mechanisms. J Neurochem 1975;24:903–907.

385 Beverly JL, Martin RJ: Effect of glucoprivation on glutamate decarboxylase activity in the ventromedial nucleus. Physiol Behav 1991;49:295–299.

386 Grandison L, Guidotti A: Stimulation of food intake by muscimol and beta-endorphin. Neuropharmacology 1977;16:533–536.

387 Kelly J, Grossman SP: GABA and hypothalamic feeding systems: a comparison of GABA, glycine, and acetylcholine agonists and their antagonists. Pharmacol Biochem Behav 1979;11:649–652.

388 Kamatchi GL, Bhakthavatsalam P, Chandra D: Inhibition of insulin hyperphagia by gamma-aminobutyric acid antagonists in rats. Life Sci 1984;34:2297–2301.

389 Morley JE: Neuropeptide regulation of appetite and weight. Endocr Rev 1987;8: 256–287.

390 Minami T, Shimizu N, Duan S, et al: Hypothalamic neuronal activity responses to 3-hydroxybutyric acid, an endogenous organic acid. Brain Res 1990;509:351–354.

391 Oomura Y, Nakamura T, Sugimori M, et al: Effect of free fatty acid on the rat lateral hypothalamic neurons. Physiol Behav 1975;14:483–486.

392 Kasser TR, Deutch A, Martin RJ: Uptake and utilization of metabolites in specific brain sites relative to feeding status. Physiol Behav 1986;36:1161–1165.

393 Kasser TR, Harris RBS, Martin RJ: Level of satiety: In vitro energy metabolism in brain during hypophagic and hyperphagic body weight recovery. Am J Physiol 1989; 257:R1322–R1327.

394 Cabanac M: Physiological role of pleasure. Science 1971;173:1103–1107.

395 Rolls BJ, Rolls ET, Rowe EA, et al: Sensory specific satiety in man. Physiol Behav 1981;27:137–142.

396 Rolls BJ, Rowe EA, Rolls ET, et al: Variety in a meal enhances food intake in man. Physiol Behav 1981;26:215–221.

397 Rolls BJ, Van Duijenvoorde PM, Rowe EA: Variety in the diet enhances intake in a meal and contributes to the development of obesity in the rat. Physiol Behav 1983; 31:21–27.

398 Drewnowski A, Grinker JA, Hirsch J: Obesity and flavor perception: Multidimensional scaling of soft drinks. Appetite 1982;3:361–368.

399 Garcia J, Hankins WG, Rusiniak KW: Behavioral regulation of the milieu interne in man and rat. Science 1974;185:824–831.

400 LeBlanc J, Cabanac M, Samson P: Reduced post-prandial heat production with gavage as compared with meal feeding in human subjects. Am J Physiol 1984;246: E95–E101.

401 Casado J, Fernandez-Lopez JA, Esteve M, et al: Rat splanchnic net oxygen consumption, energy implications. J Physiol 1990;431:557–569.

402 De Jonge L, Agoues I, Garrel DR: Decreased thermogenic response to food with intragastric vs. oral feeding. Am J Physiol 1991;260:E238–E242.

403 Niijima A: Effects of taste stimulation on the efferent activity of the autonomic nerves in the rat. Brain Res Bull 1991;26:165–167.

404 Hermann GE, Kohlermann NJ, Rogers RC: Hepatic-vagal and gustatory afferent interactions in the brainstem of the rat. J Auton Nerv Syst 1983;9:477–495.

405 Powley TL, Laughton W: Neural pathways involved in the hypothalamic integration of autonomic responses. Diabetologia 1981;20:378–387.

406 Rogers RC, Fryman DL: Direct connections between the central nucleus of the amygdala and the nucleus of the solitary tract: an electrophysiological study in the rat. J Auton Nerv Syst 1988;22:83–87.

407 Giza BK, Scott TR: Blood glucose selectively affects taste-evoked activity in rat nucleus tractus solitarii. Physiol Behav 1983;31:643–650.

408 Giza BK, Dheems RO, Van der Weele DA, et al: Glucagon administration affects taste sensitivity. Appetite 1989;12:212.

409 Yaxley S, Rolls ET, Sienkiewicz ZJ, et al: Satiety does not affect gustatory activity in the nucleus of the solitary tract of the alert monkey. Brain Res 1985;347:85–93.

410 Rolls ET: Neuronal activity related to the control of feeding; in Ritter RC, Ritter S, Barnes CD (eds): Feeding Behavior, Neural and Humoral Control. Orlando, Academic Press, 1986, pp 163–190.

Prof. W. Langhans, Institut für Nutztierwissenschaften, ETH Zürich,
CH–8092 Zürich (Switzerland)

Simopoulos AP (ed): Metabolic Control of Eating, Energy Expenditure and the Bioenergetics of Obesity. World Rev Nutr Diet. Basel, Karger, 1992, vol 70, pp 68–131

Energy Expenditure and Fuel Selection in Biological Systems: The Theory and Practice of Calculations Based on Indirect Calorimetry and Tracer Methods

M. Elia, G. Livesey

Dunn Clinical Nutrition Centre, Cambridge, and AFRC Institute of Food Research, Norwich, UK

Contents

Introduction

All living organisms use fuels to maintain their life cycles. Consequently the source and the fate of fuels used are of fundamental importance to biology, nutrition and biochemistry. A number of techniques for measuring fuel selection and utilization have been developed, but precise interpretation requires detailed understanding of the assumptions, and the optimal use of calculation procedures.

The measurement of energy expenditure in mammals originated in the late 18th century when Lavoiser noted how quickly blocks of ice were melted by heat released from animals and how this was related to the rate of oxygen consumption. Direct calorimeters are now fitted with sophisticated heat exchange systems and other equipment that take into account the heat loss due to evaporation [1]. Consideration of the stoichiometries of fuel oxidation have allowed the development of classical indirect methods of calorimetry. These methods depend on the estimation of heat production from measurements of oxygen consumption, carbon dioxide production, and sometimes the excretion of other substances, such as urinary nitrogen, methane and hydrogen.

Although indirect calorimetry is now used much more frequently than direct calorimetry, the direct method has played an important role in validating the indirect methods. In its turn, classical indirect calorimetry has been used to validate isotopic tracer methods which involve estimates of energy expenditure from CO_2 production. One of these methods is a simple isotope dilution technique which uses labelled bicarbonate [2, 3]. Another is a dual isotope technique which involves estimation of CO_2 production after administration of water labelled with both ^2H and ^{18}O [4,

5]. The loss of ^{18}O ($H_2{}^{18}O$) from the body pool occurs at a faster rate than that of ^2H ^2H_2O due to the excretion of ^{18}O in both water and carbon dioxide.

Development of the tracer techniques [2–5] has enabled estimates to be made of energy expenditure in habitual, free-living conditions. This contrasts with classical direct and indirect calorimetry which employs respiratory chambers that limit the living space. Although compact, mobile indirect calorimeters can be used 'in the field', they still limit the subject's ability to move freely. One of the advantages of classical indirect calorimetry over both direct calorimetry and the tracer methods is that an assessment can be made of the contribution of individual fuels (such as fat, carbohydrate and protein) to total energy expenditure. Such estimates are based on assumptions about both the composition of the fuels oxidized or stored in the body, and about the nature of the end products of metabolism [6].

The reliability of indirect calorimetry for estimating energy expenditure and fuel selection is strongly dependent on the use of appropriate fuel coefficients. The fuel coefficients usually adopted are general ones, but they vary from one fuel to another. Recently there has been an increasing use of nutritional regimens of unusual composition, especially with artificially compounded feeds that are administered enterally or parenterally in hospitals, or of native foods in nonindustrial developing regions. Therefore, it has been necessary to reconsider the principles of indirect calorimetry, particularly the use of fuel coefficients in different circumstances. Also, the increasing use of tracer techniques, which involve estimating CO_2 production alone, has made it necessary to consider in detail the variability of the energy equivalent of CO_2 and its predictability in different circumstances.

Principles of Indirect Calorimetry

The oxidation of an organic substrate is accompanied by consumption of oxygen and production of carbon dioxide and water (table 1). Any of these can be used to predict energy expenditure associated with the oxidation of a single fuel, although water production is difficult to measure easily and accurately. However, both oxygen consumption and CO_2 production can be measured with the precision needed to estimate heat production. To enable such estimates to be made, knowledge is needed of the

Table 1. Oxidation of carbohydrate (glucan), dioleylpalmitate, Kleiber's standard protein and alcohol[1]

Carbohydrate (RQ_{carb}, 1.0; $EeqO_2$, 21.12 kJ/l)
$$[\underset{\text{glucan}}{C_5H_{10}O_5}] + 6\ O_2 = 6\ CO_2 + 5\ H_2O;\ \Delta E_{carb},\ -2{,}840\ kJ$$

Dioleylpalmitate (RQ_{fat}, 0.71; $EeqO_2$, 19.59 kJ/l)
$$C_{55}H_{102}O_6 + 77.5\ O_2 = 55\ CO_2 + 51\ H_2O;\ \Delta E_{fat},\ -34{,}022\ kJ$$

Kleiber's standard protein[2] (RQ_{prot}, 0.833; $EeqO_{2prot}$, 19.47 kJ/l)
$$C_{100}H_{159}N_{26}O_{32}S_{0.7} + 104.0\ O_2 = 86.6\ CO_2 + 50.6\ H_2O + 11.7\ \underset{\text{urea}}{N_2H_4CO} + 1.3\ \underset{\text{aq. ammonia}}{NH_4OH}$$
$$+\ 0.43\ \underset{\text{creatinine}}{N_3C_4H_7O} + 0.7\ H_2SO_4;\ \Delta E_{prot},\ -45{,}376\ kJ$$

Alcohol (RQ_{alc}, 0.667; $EeqO_{2alc}$, 20.33 kJ/l)
$$C_2H_5OH + 3\ O_2 = 2\ CO_2 + 3\ H_2O;\ \Delta E_{alc},\ -1{,}367\ kJ$$

[1] ΔE_{carb}, ΔE_{fat}, ΔE_{prot} and ΔE_{alc} are the enthalpy changes associated with the oxidation of carbohydrate (s), fat (s), protein (s) and alcohol (l) respectively to carbon dioxide (g), water (l), ammonia (aq), urea (aq), creatinine (s) and sulphuric acid ($H_2SO_4 \cdot 115H_2O$) (s = solid; l = liquid; aq = aqueous; g = gas).
[2] End products of nitrogen metabolism urea, ammonia and creatinine in the nitrogen mass ratio of 9:5:5.

amount of heat released per litre of O_2 consumed or per litre of CO_2 produced. These amounts or coefficients are termed energy equivalents of oxygen or carbon dioxide. The coefficients for the oxidation of carbohydrate, fat, protein and alcohol are given in table 2. These values are general or standard values which are used in subsequent derivations in this paper. The energy equivalents of oxygen ($EeqO_2$) for different substrates are less variable than those for CO_2 ($EeqCO_2$). Therefore, when more than one fuel is being oxidized O_2 consumption is likely to reflect energy expenditure more accurately than is CO_2 production.

Classical indirect calorimetry is concerned with the oxidation of more than one fuel and requires for most purposes the simultaneous measurements of O_2 consumption and CO_2 production. These simultaneous measurements permit a more accurate means of predicting energy expenditure than either O_2 consumption or CO_2 production alone. The greater accuracy is achieved because the ratio of CO_2 produced to O_2 consumed indicates the contribution of different fuels to total fuel utilization. When three fuels are used, the rate of oxidation of the third substrate has to be either

Table 2. The standard values of respiratory quotient and the $EeqO_2$ and $EeqCO_2$ for carbohydrate (as glucan) fat, protein and alcohol that are used in the examples given in this paper

	RQ	$EeqO_2$, kJ/l	$EeqCO_2$, kJ/l
Carbohydrate[1]	1.000	21.12	21.12
Fat	0.710	19.61	27.62
Protein[2]	0.835	19.48	23.33
Alcohol	0.667	20.33	30.49

[1] Polyglucose (glucan).
[2] Metabolized to urea, creatinine and ammonia in the nitrogen mass ratio of 95:5:5. 1 g urinary N is associated with the utilization 5.95 litres O_2 and release of 4.968 litres of CO_2 and 116 kJ. These values take into account the calorimetric coefficients of nitrogenous substances (other than urea, creatinine and ammonia such as amines, uric acid, etc.) which account for about 5% urinary N in normal man [see 7].

assumed or estimated, e.g. the loss of nitrogen in urine can be used to indicate protein oxidation. The problem becomes more complex when an increasing number of fuel is used in the oxidation mixture, and when the end products of metabolism are substances other than CO_2, water and urinary nitrogen (e.g. production of hydrogen, methane and ketone bodies). These complexities can be appreciated by considering the variabilities in the calorimetric coefficients for fat, carbohydrate, protein and other substances ($EeqO_2$, $EeqCO_2$, RQ), and how these influence the calculation of energy expenditure.

Establishing the Calorimetric Coefficients for Fat, Carbohydrate and Protein

The coefficients that need to be established for the purposes of indirect calorimetry include the respiratory quotients (RQ), and the energy equivalents of oxygen ($EeqO_2$), and carbon dioxide ($EeqCO_2$).

Fat
The calorimetric coefficients for the oxidation of fat can be calculated from three parameters: their heats of combustion, O_2 consumption and CO_2 production.

Heats of combustion of triglycerides are determined using a bomb calorimeter [8, 9], but they can also be predicted from the fatty acid composition [7, 10]. The latter method is particularly useful because not all types of fats and fatty acids have been prepared in sufficient quantity or purity for bomb calorimetry. Thus, using the regularities of the heat of combustion of fatty acids of different chain length and different degree of saturation, Livesey [10] established the following equation:

Heat of combustion of fatty acids (kJ/mol) = $-(653n-166d-421)$ (1)

where n is the number of carbon atoms/mol fatty acid, and d is the number of double bonds per fatty acid. The overall negative sign indicates that heat is released to the environment. If the fatty acid composition of a triglyceride is known, the combined heats of combustion of fatty acids can be calculated from the above formula and added to the heat combustion of glycerol ($-1,661$ kJ/mol) [11], to give the heat of combustion of the triglyceride. The heat of hydrolysis of the fatty acyl-glycerol ester bonds (3/mol triglyceride) is considered negligible ($<0.3\%$) compared to the total heat of combustion of the triglyceride [7].

The gaseous exchange (O_2 consumption, CO_2 production) associated with the oxidation of a chemically defined triglyceride is predictable, but the precise values vary with the fatty acid composition [7]. Table 1 shows the oxidative stoichiometry for dioleylpalmitate ($C_{55}H_{102}O_6$).

The calorimetric coefficients for triglyceride from human adipose tissue have been found to show little variability irrespective of the subject's age or country of domicile (table 3). The corresponding values for dietary fats and oils are indicated in table 4. In general, fats with high RQs also have a high $EeqO_{2\,fat}$ and low $EeqCO_{2\,fat}$. The ranges of values for the $EeqCO_{2\,fat}$, $EeqO_{2\,fat}$ and RQ of adipose tissue fat and for most dietary fats are remarkably small (table 4). The influence of atypical dietary fatty acids in mixed diets on the accuracy of indirect calorimetry when using general coefficients is 'buffered' by the presence of other more conventional fatty acids. Furthermore, fats which have a relatively high RQ and $EeqO_{2\,fat}$ (and a low $EeqCO_{2\,fat}$), for example medium chain triglyceride, do not usually constitute a large proportion of the dietary fat intake. The most common RQ for dietary fat from conventional sources is about 0.71, the $EeqO_{2\,fat}$ is about 19.6 kJ/l, and the $EeqCO_{2\,fat}$ is about 27.6 kJ/l (fig. 1) (the energy equivalent of a gas refers to the heat produced or energy expended per litre gas and for the purposes of this paper it is assigned a positive value).

Table 3. Respiratory quotients, energy equivalents of O_2 and CO_2, and heats of combustion of human adipose tissue triglyceride

Adipose tissue triglyceride	RQ	Energy equivalent of gas		Heat of combustion
		kJ/l O_2 (kcal/l O_2)	kJ/l CO_2 (kcal/l CO_2)	kJ/g (kcal/g)
Adipose tissue fat				
Pre-term infants, USA	0.7095	19.58 (4.680)	27.60 (6.597)	39.51 (9.443)
Full-term infants, USA	0.7103	19.60 (4.685)	27.59 (6.595)	39.46 (9.431)
Infants < 2 y fed MCT, UK	0.7105	19.59 (4.682)	27.57 (6.590)	39.47 (9.434)
Infants < 2 y, UK	0.7109	19.59 (4.682)	27.56 (6.590)	39.47 (9.434)
Children 2 y, FRG	0.7112	19.60 (4.685)	27.56 (6.595)	39.50 (9.441)
Adults, USA	0.7120	19.60 (4.685)	27.53 (6.579)	39.56 (9.455)
Adults, USA	0.7101	19.59 (4.682)	27.58 (6.594)	39.57 (9.457)
Adults, UK	0.7100	19.59 (4.682)	27.58 (6.590)	39.60 (9.465)
Adults, Japan	0.7140	19.63 (4.692)	27.49 (6.571)	39.53 (9.448)
Adults, Israel	0.7145	19.62 (4.689)	27.46 (6.563)	39.56 (9.455)
Adults[1]	0.713	20.06 (4.794)	28.13 (6.724)	39.91 (9.539)
Adults, UK[1]	0.711	19.87 (4.749)	27.95 (6.679)	39.77 (9.505)

[1] Determined directly by bomb calorimetry and elemental analysis; all other data were calculated from fatty acid composition data. With permission from Am J Clin Nutr [7].

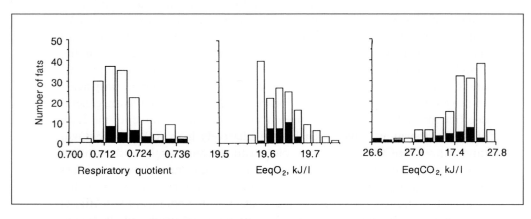

Fig. 1. Distribution of the RQ, $EeqO_2$ and $EeqCO_2$ for 125 conventional food fats and 28 artificial feeds. Fats with a higher RQ tend to have a higher $EeqO_2$ and lower $EeqCO_2$ (calculated from compositional data [41]).

Table 4. Energy equivalents of CO_2 and O_2 and respiratory quotient of various fats and oils[1]

	EeqCO$_2$, kJ/l (kcal/l)	EeqO$_2$, kJ/l (kcal/l)	RQ
Fats and oils in:			
Conventional foods	26.78–27.1	19.56–19.74	0.706–0.737
	(6.40–6.61)	(4.68–4.72)	
Artificial foods	26.68–27.62	19.58–19.66	0.709–0.737
	(6.38–6.60)	(4.68–4.70)	
Fat in:			
British household food supply	27.51 (6.58)	19.59 (4.68)	0.712
Vegan diet	27.44 (6.56)	19.62 (4.69)	0.715
Omnivore diet	27.55 (6.59)	19.59 (4.68)	0.711
Safflower seed oil	27.24 (6.51)	19.67 (4.70)	0.722
Cod fish oil	26.78 (6.40)	19.74 (4.72)	0.737
Human milk (mature)	27.58 (6.59)	19.58 (4.68)	0.710
High corn oil diet	27.38 (6.54)	19.63 (4.69)	0.717
Infant formula diets	27.13–27.58	19.59–19.67	0.713–0.722
	(6.48–6.59)	(4.68–4.70)	
Atypical fats			
Medium chain triglyceride	26.64 (6.37)	19.69 (4.71)	0.739
Monobutyrin ($C_7H_{14}O_4$)	24.55 (5.87)	20.23 (4.83)	0.824
Monoacetoacetin ($C_7H_{12}O_5$)	21.90 (5.23)	20.43 (4.88)	0.933
Human adipose tissue			
Adults	27.44–27.61	19.60–19.62	0.710–0.715
	(6.56–6.60)	(4.68–4.69)	
Preterm infants	27.60 (6.60)	19.58 (4.68)	0.709

[1] Calculated using compositional data [7]. With permission from Am J Physiol [6].

An analysis of the fats in 36 artificial feeds used for enteral nutrition in humans [7] revealed a range of calorimetric coefficients similar to those noted for conventional food values. However, the most common RQ$_{fat}$ was 0.72 (c.f. 0.71 for conventional fats), and the most common values for EeqO$_{2\,fat}$ and EeqCO$_{2\,fat}$ were found to be about 19.66 kJ/l (c.f. 19.6) and 27.31 kJ/l (c.f. 27.6) respectively [7]. A further difference, which applies also to carbohydrates and amino acids, is that those nutrients in artificial feeds are commonly ingested to the exclusion of all other sources of nutri-

Table 5. The respiratory quotients, heats of combustion and the energy-gas equivalents of carbohydrates

	RQ	Heat of combustion	Energy-gas equivalents	
			O_2	CO_2
		kJ/g (kcal/g)	kJ/l (kcal/l)	kJ/l (kcal/l)
Starch	1.000	−17.48 (−4.178)	21.12 (5.048)	21.12 (5.048)
Glycogen	1.000	−17.52 (−4.187)	21.12 (5.048)	21.12 (5.048)
Sucrose	1.000	−16.48 (−3.939)	20.97 (5.012)	20.97 (5.012)
Maltose	1.000	−16.49 (−3.941)	20.99 (5.017)	20.99 (5.017)
Lactose	1.000	−16.50 (−3.944)	21.00 (5.019)	21.00 (5.019)
Glucose	1.000	−15.56 (−3.719)	20.84 (4.981)	20.84 (4.981)
Galactose	1.000	−15.56 (−3.719)	20.84 (4.981)	20.84 (4.981)
Fructose	1.000	−15.61 (−3.731)	20.91 (4.997)	20.91 (4.997)
Glycerol	0.857	−18.03 (−4.305)	21.16 (5.057)	24.69 (5.901)
Xylitol	0.909	−16.96 (−3.989)	19.92 (4.701)	21.91 (5.237)
Sorbitol	0.920	−16.71 (−3.993)	20.89 (4.993)	22.71 (5.427)
Maltitol	0.960	−16.98 (−4.058)	20.87 (4.988)	21.74 (5.196)
Lactitol	0.960	−16.98 (−4.058)	20.87 (4.988)	21.74 (5.196)

tion. Consequently, specific calorimetric coefficients are more likely to apply to artificial feeds than to conventional diets. This topic is discussed later in relation to errors arising from the use of indirect calorimetry (see last two sections).

Carbohydrate

The calorimetric coefficients for various carbohydrates are shown in table 5. For many carbohydrates, including starch, glycogen, sucrose, maltose, lactose, glucose and galactose, the values for RQ_{carb} are identical. However, the values for $EeqO_2$ and $EeqCO_{2\,carb}$ differ slightly because of the heat released during the hydrolysis of those carbohydrates with glycosidic linkages. The values of $EeqCO_2$ for disaccharides (maltose, sucrose, lactose) are therefore about 0.5% lower than for polysaccharides (starch and glycogen), and those for the monosaccharides are about 1% lower than for the polysaccharides. The $EeqO_2$ of glucose is 1.3% lower than the value for glycogen and starch. After an overnight fast, when indirect calorimetry is often used, the value of 21.12 kJ/l is appropriate since glycogen is the source of carbohydrate energy. However, since about 35% of the carbohydrate in Western diets is in the form of simple sugars, the $EeqO_2$ of these diets is slightly lower than 21.12 kJ/l. These differences are usually so small ($\leq 0.2\%$ of total energy expenditure), that for practical purposes they can often be ignored.

Larger differences exist between glycogen or starch and the sugar alcohols (table 5). The latter are usually minor dietary constituents [12], but in some situations they may become significant sources of dietary energy. For example, sorbitol and xylitol have been used as the major carbohydrates in artificial parenteral feeds [13]. Free glycerol is also used as a emulsifying agent in lipid emulsions used in parenteral nutrition (e.g. 22.5 g/l of 10 or 20% Intralipid Kabi Vitrum, Stockholm, Sweden). The errors associated with the use of standard equations to calculate energy expenditure and fuel selection when atypical fuels are oxidized are considered later.

Protein
Establishing calorimetric coefficients for the metabolism of protein has proved to be difficult and not as precise as for fat and carbohydrate. Unlike fat and carbohydrate, protein is always only partially oxidized to nitrogenous end products (e.g. urea). It is obvious that the heat released when a protein is partially oxidized to CO_2, H_2O and urea (heat of combustion of urea is 22.6 kJ/gN) is less than when combustion is complete. Another difficulty is the variety of end products of protein metabolism, the proportion of which may differ widely between species, and to a lesser extent within the same species under different metabolic circumstances. Such variety leads to a range of different values for both $EeqO_{2\,prot}$ and RQ_{prot}. Tables 6 and 7 show the coefficients of Kleiber's standard protein [14] when it is oxidized to urea, ammonia, creatinine and allantoin. For completeness, creatinine is included in the tables, although it is not normally the dominant nitrogenous end product in any species. Depending on the nitrogenous end products, the possible values for the RQ of Kleiber's standard protein range from 0.707 to 0.95, those for $EeqO_{2\,prot}$ range from 19.11 to 19.68 kJ/l, those for $EeqCO_2$ range from 20.72 to 27.60 kJ/l (tables 6, 7).

The calorimetric coefficients for protein given by Lusk [15], Loewy [16] and Magnus-Levy [17] have had enormous use in calculating fuel selection from gaseous exchange, and urine N excretion. The Lusk/Loewy factors have been adopted by a large number of scientists including Weir [18], Abramson [19], and McGilvery [20], and all or some of the Magnus-Levy factors [17] have been adopted by a large number of other workers

Table 6. The conversion of 'Kleiber's standard protein' to various nitrogenous end products

---

Protein to urea
$$C_{100}H_{159}N_{26}O_{32}S_{0.7} + 105.3\ O_2 = 13\ CON_2H_4 + 87\ CO_2 + 52.8\ H_2O + 0.7\ H_2SO_4;$$
$$\Delta E,\ -45,950\ kJ;\ RQ,\ 0.826$$

Protein to ammonia
$$C_{100}H_{159}N_{26}O_{32}S_{0.7} + 105.3\ O_2 = 26\ NH_4OH + 100\ CO_2 + 13.8\ H_2O + 0.7\ H_2SO_4;$$
$$\Delta E,\ -46,450\ kJ;\ RQ,\ 0.95$$

Protein to creatinine
$$C_{100}H_{159}N_{26}O_{32}S_{0.7} + 79.3\ O_2 = 8.667\ C_4H_7ON_3 + 65.332\ CO_2 + 48.466\ H_2O +$$
$$0.7\ H_2SO_4;\ \Delta E,\ -33,960\ kJ;\ RQ,\ 0.824$$

Protein to uric acid
$$C_{100}H_{159}N_{26}O_{32}S_{0.7} + 95.5\ O_2 = 6.5\ C_5H_4O_3N_4 + 67.5\ CO_2 + 65\ H_2O + 0.7\ H_2SO_4;$$
$$\Delta E,\ -41,880\ kJ;\ RQ,\ 0.707$$

Protein to allantoin
$$C_{100}H_{159}N_{26}O_{32}S_{0.7} + 98.8\ O_2 = 6.5\ C_4H_6O_3N_4 + 74\ CO_2 + 59.3\ H_2O + 0.7\ H_2SO_4;$$
$$\Delta E,\ -43,254\ kJ;\ RQ,\ 0.749$$

ΔE = Change in enthalpy. With permission from Am J Physiol [6].

---

including Consolazio et al. [21], Passmore and Eastwood [22], Ben-Porat et al. [23] and Frayn [24]. Indeed many of the currently used formulae for calculating energy expenditure and fuel selection originate from these sources. However, the traditional procedures for calculating the protein coefficients for use in calorimetry have problems.

The methods for establishing calorimetric coefficients for protein has long been controversial. Zuntz [25] appears to have been the first to suggest calorimetric coefficients for protein metabolism. Using data from the metabolic balance studies undertaken by Rubner, Zuntz [25] calculated the amounts of H, N, C, O (and apparently not S) of dietary protein used for oxidative metabolism at zero nitrogen balance (sulphur balance was ignored). All the above elements in urine were assumed to arise as end products of protein metabolism. All the elements in faeces, after discounting those in faecal fat, were assumed to arise from dietary or endogenous protein. Unrecovered H, C and O were assumed to be oxidized and the amounts of oxygen required and carbon dioxide produced in this process were calculated. This approach erroneously assumed that all the substrates

Table 7. The respiratory quotients and the energy equivalents of O_2 (EeqO$_{2prot}$) and CO_2 (EeqCO$_{2prot}$) for the oxidation of Kleiber's standard protein to different nitrogenous end products

Conversion[1]	RQ	EeqO$_{2prot}$, kJ/l	EeqCO$_{2prot}$, kJ/l
Protein to urea	0.826	19.48	23.58
Protein to ammonia	0.950	19.68	20.72
Protein to creatinine	0.824	19.11	23.19
Protein to uric acid	0.707	19.57	27.68
Protein to allantoin	0.749	19.54	25.08

[1] For equations see table 6.

in urine and all organic matter in faeces other than fat, are products of protein metabolism. Thus the calculations assume that there is no faecal carbohydrate. The coefficients obtained by Zuntz [25] were remarkably good considering both the inappropriateness of the assumptions and the inaccuracies of balance studies. Magnus-Levy [17], who also used the elemental balance procedure, calculated the calorimetric coefficient for protein (RQ$_{prot}$, 0.809; EeqO$_{2\,prot}$, 19.25 kJ/l) as the mean values for Zuntz's 'muscle substance', and casein. Since the values for RQ and EeqO$_{2\,prot}$ given by Zuntz [25] were 0.793 and 18.73 kJ/l O_2 respectively, we calculate that the corresponding values for casein obtained by Magnus-Levy must have been 0.825 and 19.76 kJ/l.

The most commonly cited reference to a balance study producing calorimetric coefficients for protein is probably that of Lusk [15] (table 8). Lusk quotes Loewy [16] as the source of his information. The original values of Loewy were modified between the third (1917) and fourth edition (1928) of Lusk's book, to take into account new information on the densities of O_2 and CO_2. This changed Loewy's estimates of the RQ$_{prot}$ from 0.801 to 0.8016 and of EeqO$_{2\,prot}$ from 18.765 to 18.67 kJ/l. Loewy's estimates, which were derived from elemental balance measurements made on a single dog (table 8), were similar to those obtained by Zuntz [25].

It should also be emphasized that it is a mistake to use the food energy values for protein (4 kcal (16.74 kJ)/g) and fat (9 kcal (37.66 kJ)/g) when calculating metabolic rate from gaseous exchanges and urine N [21, 22, 27–29]. After accounting for the end products of oxidation (e.g. urea) these

Table 8. Derivation of the constants associated with protein oxidation according to the Loewy procedure[1]

		Grams of element				
		C	H	O	N	S
1	Gain from 100 g meat protein	52.38	7.27	22.68	16.65	1.02
2	Loss to urine	9.406	2.663	14.099	16.28	1.02
3	Loss to faeces	1.471	0.212	0.889	0.37	–
4	Residuum used for respiration (1–2–3)	41.50	4.40	7.69		
5	Intramolecular water		0.961	7.69		
6	Deduct intramolecular water[2] (4–5)	41.5	3.439			

[1] Loewy [16] modified by Lusk [15]. See also Peters and van Slyke [26].
[2] These quantities of C and H require 138.18 g O_2 (96.70 litres) to produce 152.17 g CO_2 (77.52 litres). The RQ for meat protein is 77.52/96.70 or 0.8016. This means that 5.941 litres O_2 (96.7/16.28) are used in oxidation for every gN excreted in urine. If the metabolic energy value of 1 g of available meat protein is 18.058 kJ (4.316 kcal) [15] then the $EeqO_{2prot}$ is 18.058/0.967 or 18.674 kJ (4.463 kcal).

substances release the same amount of heat and CO_2 and use the same amount of oxygen in vivo as they do in a bomb calorimeter. It is clear that the heat of combustion values should be used for fat and carbohydrate. For protein a value intermediate between its heat of combustion (generally 23.6 kJ/g [10]), and its food energy value (generally 16.7 kJ/g [30]) is appropriate. The value is the difference between the heats of combustion of protein and its nitrogenous end products of metabolism – see also below for heat of solution and dilution of the end products). The energy expended relates the amount of O_2 consumed and CO_2 produced, and is independent of the amount of nutrients absorbed from the gut.

Apart from the difficulties associated with ascribing faecal constituents to products of protein metabolism, the elemental balance methods assume all urinary constituents to be end products of protein metabolism. However, urine contains small amounts of protein and amino acids, as well as substances derived from fat and carbohydrate metabolism [e.g. 31]. Furthermore, some nitrogenous products in urine are derived from the combined metabolism of protein, and carbohydrate or fat. Yet another difficulty arises from the fact that nitrogenous substances such as urea absorb heat while dissolving in water (i.e. solvation is endothermic), whilst others such as ammonia release heat (i.e. solvation is exothermic). Adjustments to the physiological state of the end products, rather than the standard state after bomb calorimetry, requires knowledge of the urinary constituents and their associated heat of solution and dilution. Related to this is a problem of the oxidation of sulphur in protein to sulphuric acid and the heat of solution and dilution of this acid.

An alternative to the elemental balance method to determine the calorimetric coefficients of protein (RQ, $EeqO_2$, $EeqCO_2$), is the stoichiometric approach which involves calculating the amount of O_2 utilized and CO_2 produced during the oxidation of a protein of known elemental composition to specified nitrogenous end products in solution. Examples are shown in table 6. The heat exchange needs to be calculated also; this requires knowledge of the heat of combustion (or heats of formation) of reactants and products, the state in which reactants and products exist, and information on their heats of solution and dilution. To calculate the $EeqO_{2\,prot}$ and $EeqCO_{2\,prot}$ (tables 6, 7) it is also necessary to have information about the physiological end products of sulphur metabolism (dilute aqueous sulphuric acid). However, it should be noted that tabulated heats of combustion for organic sulphur compounds refer to a variety of end products such as sulphur [32], sulphur trioxide [33] and dilute sulphuric acid [8]. The last is the more physiological.

Table 9 gives values for the heat of combustion of amino acids and is derived from the critical analysis of Domalski [8]. Purity of amino acids used in the determinations is uncertain. Examination of the several published heats of combustion data for individual amino acids reveals a small degree of uncertainty [e.g. 7, 9]. Coefficients of variation between laboratories are as low as $\leq 0.1\%$ for phenylalanine, leucine and isoleucine, 0.2% for tyrosine, 0.3% for aspartic acid and asparagine, and 0.5% for glutamic acid and glycine. The accuracy of heats of combustion data should therefore be regarded as no better than 0.1–0.5%. The higher precision given in table 9 is to prevent rounding errors when values are used for calculation purposes. Reported heats of combustion for threonine vary widely, with a coefficient of variation of 2%. The tabulated value for threonine (table 9) is consistent with that for serine and the heat increment associated with the replacement of hydrogen in serine with the methyl group in threonine. Reported heats of combustion data for methionine (and cysteine) vary widely depending on the state of the end products of sulphur oxidation (S_2,

Table 9. Heats of combustion and metabolizable energy values of amino acids (AA)[1]

AA	Mol. wt.	N/ mol	Heat of combustion, kJ per:					Metabolizable energy[1], kJ per:				
			mol	g AA	g P[2]	gN AA	gN P[2]	mol	g AA	g P[2]	gN AA	gN P[2]
Ala	89.1	1	−1,620	−18.18	−23.07	−115.7	−117.1	−1,296	−14.55	−18.52	−92.6	−94.0
Asp	132.1	1	−1,602	−12.13	−14.22	−114.4	−115.9	−1,278	−9.68	−11.38	−91.3	−92.7
Asn	133.1	2	−1,928	−14.49	−16.92	−68.9	−69.9	−1,281	−9.62	−11.30	−45.7	−46.5
Glu	147.1	1	−2,244	−15.25	−17.54	−160.3	−161.7	−1,920	−13.06	−15.03	−137.2	−139.6
Gln	146.2	2	−2,571	−17.59	−20.21	−91.8	−92.5	−1,924	−13.16	−15.16	−68.7	−69.4
Gly	75.1	1	−974	−12.97	−17.41	−69.6	−71.0	−650	−8.66	−11.74	−46.5	−47.9
Pro	115.1	1	−2,727	−23.69	−28.29	−194.8	−196.2	−2,403	−20.88	−24.96	−171.7	−173.1
Ser	105.1	1	−1,455	−13.84	−16.93	−103.9	−105.4	−1,131	−10.77	−13.22	−80.8	−82.2
Cys	121.2	1	−2,261	−18.66	−22.10	−161.5	−162.9	−1,937	−15.99	−18.87	−138.4	−139.8
Tyr	181.2	1	−4,430	−24.45	−27.27	−316.4	−317.9	−4,106	−22.66	−25.28	−293.3	−294.7
Orn	132.2	2	−3,030	−22.92	−26.71	−108.2	−108.9	−2,383	−18.03	−21.04	−85.1	−85.8
Iso	131.2	1	−3,584	−27.32	−31.84	−256.0	−257.4	−3,260	−24.85	−28.98	−232.9	−234.3
Leu	131.2	1	−3,583	−27.31	−31.83	−255.9	−257.4	−3,259	−24.84	−28.97	−232.8	−234.2
Lys	146.2	2	−3,683	−25.19	−28.88	−131.5	−132.3	−3,036	−20.77	−23.84	−108.4	−109.1
Met	149.2	1	−3,387	−22.69	−25.97	−241.9	−243.4	−3,062	−20.53	−23.50	−218.7	−220.2
Phe	165.2	1	−4,645	−28.12	−31.69	−331.8	−333.2	−4,321	−26.16	−29.49	−308.7	−310.1
Thr	119.1	1	−2,101	−17.64	−20.98	−150.1	−151.5	−1,777	−14.92	−17.87	−127.0	−128.4
Try	204.2	2	−5,629	−27.57	−30.34	−201.0	−201.8	−4,982	−24.40	−26.86	−177.9	−178.6
Val	117.2	1	−2,920	−24.91	−29.64	−208.6	−210.0	−2,596	−22.15	−26.38	−185.5	−186.9
Arg	174.2	4	−3,738	−21.46	−24.06	−66.8	−67.1	−2,444	−14.03	−15.77	−43.6	−44.0
His	155.2	3	−3,259	−21.00	−23.90	−77.6	−78.1	−2,288	−14.74	−16.82	−54.5	−55.0

[1] Metabolizable energy assumes urea is the nitrogenous end product (heat of combustion −23.11 kJ/gN), and in the case of peptides adjustments are made for both water of condensation 18 g/mol AA and heat of hydrolysis, −20 kJ/mol AA (release of heat). The metabolizability of the amino acid is assumed to be 1.0.
[2] P = peptide, as a constituent of protein.

SO_3 or H_2SO_4). Combustion products for all amino acids in table 9 are: H_2O (liquid), CO_2 (gas), N_2 (gas) and $H_2SO_4 \cdot 115H_2O$ (liquid) at 25 °C and 760 mm Hg pressure. The authors are unaware of data on the heat of combustion of histidine. Therefore, a value for histidine (imidazolylalanine) was calculated from the sum of the heats of combustion of imidazole and alanine less the heat increment for the hydrogen-carbon bonds eliminated on fusion of the two compounds. Generally, it is assumed that *D* and *L* amino acids have very similar heats of combustion (i.e. within 0.5% of each other) but data on the heats of combustion of *D* and *L* amino acids are

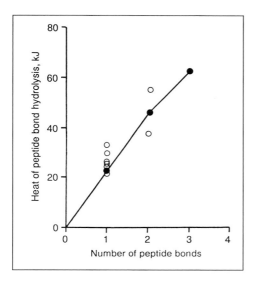

Fig. 2. The relationship between the number of peptide bonds and the heat released during hydrolysis of the peptides. ● = Peptides of glycine; ○ = other peptides.

either too sparse or too inaccurate to make an assessment of possible small differences between them.

The heats of combustion of the amino acids have been expressed in different ways by previous workers (see below) and this may have led to some confusion. In table 9 they are expressed in units of kJ either per mol amino acids, or per g amino acid, or per g amino acid residue (as protein) or per g amino acid nitrogen. The values expressed per g protein take cognizance both of the loss of weight due to the water of condensation (18 g/mol amino acid) and of the additional heat arising from the hydrolysis of the peptide bond. Precise values for the heat of hydrolysis of peptide bonds appear to depend on whether or not an account is made of the state (solid or aqueous) in which the peptides and products occur, and whether those in solution are fully charged. Suggested values for dipeptide hydrolysis are −5 to −10 kJ/peptide bond [9]. A higher value of −15 kJ/peptide bond is calculated from information supplied by Shulz [34]. The average value given by Rawitscher et al. [35] is −8 kJ/peptide bond. However, a more negative value is expected based on the differences in the heats of combustion of peptides and their constituent amino acids. Figure 2 shows values

Table 10. Heats of combustion, metabolizable energy values (ME) energy equivalent of O_2 (EeqO$_2$), respiratory quotients (RQ) and heats of solution of amino acids, and the effects of changing the end products of metabolism

Amino acids and end products of metabolism	Molecular weight	Heat of combustion[1]		ME[2] kJ/mol	EeqO$_2$[2] kJ/l O$_2$	O$_2$[2] mol/mol	RQ[2] CO$_2$/O$_2$	Heat of solution and dilution[3] kJ/mol
		kJ/gN	kJ/mol					
Amino acids								
Alanine	89.1	−115.7	−1,620	−1,296	19.27	3.0	0.833	+8
Aspartic acid	133.1	−114.4	−1,620	−1,278	19.01	3.0	1.167	+26
Asparagine	152.1	−68.9	−1,928	−1,281	19.05	3.0	1.000	+24
Glutamic acid	147.1	−160.3	−2,244	−1,920	19.04	4.5	1.000	+24
Glutamine	146.2	−91.3	−2,581	−1,924	19.07	4.5	0.889	+21
Glycine	75.1	−69.6	−974	−650	19.33	1.5	1.000	+14
Proline	155.2	−194.8	−2,727	−2,403	19.49	5.5	0.818	0
Serine	105.1	−103.9	−1,455	−1,131	20.18	2.5	1.000	+21
Cysteine	121.2	−161.5	−2,261	−1,937	19.20	4.5	0.556	
Tyrosine	181.2	−316.4	−4,430	−4,106	19.28	9.5	0.895	+25
Ornithine	132.2	−108.2	−3,030	−2,383	19.33	5.5	0.727	
Isoleucine	131.2	−256.0	−3,584	−3,260	19.39	7.5	0.733	+4
Leucine	131.2	−255.9	−3,583	−3,259	19.38	7.5	0.733	+4
Lysine	146.2	−131.5	−3,683	−3,036	19.35	7.0	0.714	−16
Methionine	149.2	−241.9	−3,387	−3,062	18.21	7.5	0.600	+12
Phenylalanine	165.2	−331.8	−4,645	−4,321	19.27	10.0	0.850	+12
Threonine	119.1	−150.1	−2,161	−1,777	19.82	4.0	0.875	
Tryptophane	204.2	−201.0	−5,629	−4,982	19.33	11.5	0.870	+6
Valine	117.2	−208.6	−2,920	−2,596	19.30	6.0	0.750	+4
Arganine	174.2	−66.8	−3,738	−2,444	19.83	5.5	0.727	
Histidine	155.2	−77.6	−3,259	−2,288	20.42	5.0	0.900	+14
End products								
Urea	60.1	−22.6	−632		19.24	1.50	0.666	+15
Ammonia	17.0	−27.3	382		20.70	0.75	0.000	−29
Creatinine	113.1	−55.6	−2,337		19.85	5.25	0.762	−
Creatine	131.1	−55.3	−2,324		19.75	5.25	0.762	−
Uric acid	158.1	−34.3	−1,921		19.04	4.50	1.111	−

[1] In the standard state the end products are N_2, CO_2, H_2O and $H_2SO_4 \cdot 115 H_2O$.

[2] These values for amino acids assume urea is the sole nitrogenous end product of metabolism.

[3] The values are approximate due to uncertainties about purity of amino acids, the activity coefficients and solubility data used to estimate heats of solution and dilution [36]. The exception is the value for glutamine which is calculated from Wilhout [37]. The negative value for the heat of solution and dilution as well as for ME/heat of combustion indicates energy released (exothermic or energy loss from reactants), whilst the positive values for urea indicates energy uptake (endothermic – or energy gain by reactants).

calculated for di-, tri- and tetrapeptides using the heat of combustion data of Domalski [8]. On average the extra heat released is 20 ($\pm$ 3 SD) kJ/ peptide bond. This value has been selected for the present purposes although it is uncertain whether the value changes with longer peptides or proteins.

The metabolizable energy values of the amino acids are also shown in table 10, again expressed in several different ways. The values were based on the end products of metabolism being H_2O (liquid), CO_2 (gas), urea (aqueous), CO_2 (gas) and $H_2SO_4 \cdot 115H_2O$ (liquid). The difference between metabolizable energy and heat of combustion (see table 10) is due to loss of aqueous urea (or ammonia) in urine rather than gaseous nitrogen. When 5% of the nitrogen loss is ammonia, the values for metabolizable energy values are 0.1 kJ/g amino acid nitrogen smaller than those given in table 10. For the oxidation of protein in man the major nitrogenous end products are urea, creatinine and ammonia, which are usually excreted in a nitrogen mass ratio of about 90:5:5 respectively. Oxidation of protein or amino acid mixtures to urea, ammonia and creatinine in this nitrogenous ratio produces values of metabolizable energy that are about 1.6 kJ/g amino acid nitrogen smaller than when urea is the only nitrogenous end product of protein metabolism (table 10).

Values of RQ and $EeqO_2$ for various amino acid mixtures used in artificial nutrition are shown in table 11 and for selected conventional food proteins in table 12. Values of RQ range from 0.752 (or 0.738 for branded chain mixtures) to 0.841 while values of $EeqO_2$ range from 19.27 to 19.46 kJ/g. For food protein there is some uncertainty about the proportion of glutamine:glutamic acid and of asparagine:aspartic acid, but this does not affect the overall $EeqO_{2\,prot}$ since glutamine and glutamate have virtually the same $EeqO_2$, and so do asparagine and aspartic acid.

The values of $EeqO_2$ for most amino acids, and consequently for most proteins, are higher than those used by Zuntz and Loewy/Lusk (see above). This is apparent irrespective of whether $EeqO_2$ is calculated for individual nitrogenous end products (e.g. urea or ammonia) or a combination of urea, ammonia and creatinine in the nitrogen mass ratios of 95:5:5 (table 2). The values for $EeqO_{2\,prot}$ given by Brouwer [38] (19.24 kJ/l) and Erwin et al. [39] (19.66 kJ/l) are similar to that suggested in this paper. However, that given by Brody [40] is much higher (20.19 kJ/l). The stoichiometric approach also suggests that the RQ of protein is higher than that accepted by many previous workers.

Table 11. The respiratory quotients (RQ), energy equivalents of oxygen ($EeqO_2$) and carbon dioxide ($EeqCO_2$) for the oxidation of some commercially available amino acid mixtures[1]

Amino acid mixture	Supplier	RQ	$EeqO_2$ kJ/l	$EeqCO_2$ kJ/l
Enteral				
Albumaid XI	Scientific Hospital Supplies, Liverpool, UK	0.823	19.35	23.35
Amin-Aid	Boots, Nottingham, UK	0.752	19.36	25.53
Dialamine	Scientific Hospital Supplies	0.768	19.36	25.21
Elemental 028	Scientific Hospital Supplies	0.816	19.36	23.77
Formula HF(2)	Cow & Gate, Trowbridge, UK	0.827	19.29	23.33
Hepatamine	Scientific Hospital Supplies	0.776	19.46	25.07
Maxamaid XI	Scientific Hospital Supplies	0.819	19.34	23.71
Nefranutrin	Geistlich, Chester, UK	0.748	19.27	25.77
Vivonex standard	Norwich-Eaton, Woking, UK	0.823	19.32	23.47
Vivonex HN	Norwich Eaton	0.840	19.30	22.96
Parenteral				
Aminoplasmal L10	Braun, Melsungen, FRG	0.801	19.34	24.13
Aminoplasmal ped	Braun	0.841	19.30	22.95
Aminoplex 14	Geistlich	0.786	19.29	24.56
Aminoplex 5 & 12	Geistlich	0.790	19.34	24.48
Aminofusin	Merck, Alton, UK	0.835	19.32	23.13
Freamin II	Boots	0.799	19.38	24.25
HepatAmine	Boots	0.787	19.46	24.71
Nephramine	McGraw, Irvine, Calif., USA	0.752	19.21	25.56
Perifusin	Merck	0.835	19.32	23.12
Synthamin (old)	Travenol, Thetford, UK	0.807	19.35	23.99
Synthamin (new)	Travenol	0.798	19.42	24.33
Travenol BCAA mix.	Travenol	0.738	19.37	26.24
Vamin 9	Kabi Vitrum, Uxbridge, UK	0.828	19.29	23.42
Vamin 14	Kabi Vitrum	0.803	19.40	24.15

[1] All data given are for urea as the nitrogenous end product.

Derivation of Equations Used to Calculate Energy Expenditure and Fuel Selection

The calculations of energy expenditure and fuel selection from gaseous exchange depend on the values of RQ and $EeqO_2$ for each substrate (fat, carbohydrate and protein) used in the calculations. Incorrect coefficients produce incorrect results. The quantitative aspects of such errors are most

Table 12. The respiratory quotients (RQ), energy equivalents of oxygen ($EeqO_2$) and carbon dioxide ($EeqCO_2$) of proteins from some common foods[1]

Protein source	RQ	$EeqO_2$, kJ/l
White (wheat) flour	0.842	19.49
Soya flour	0.835	19.53
Cows' milk	0.825	19.50
Human milk	0.822	19.52
Whole egg	0.821	19.53
Beef	0.824	19.53
Pork	0.824	19.53
Chicken	0.824	19.52
Fish	0.820	19.52
Crustacea	0.825	19.52
Asparagus	0.851	19.50
Broad beans	0.835	19.56
Haricot beans	0.835	19.54
Peas	0.831	19.55
Potatoes	0.845	19.51
Apples	0.855	19.53
Dates	0.833	19.53
Almonds	0.841	19.53
Peanuts	0.841	19.55
Mean for 101 food proteins	0.833	19.53

The molar ratio of glutamine to glutamic acid and asparagine to aspartatic acid was assumed to be 1:1. When all this N is in the amine form (aspartic and glutamic acids) the $EeqO_2$ remains essentially unchanged but the RQ increases by approximately 0.01. When all this N is in glutamine and asparagine the $EeqO_2$ is again essentially unchanged but the RQ decreases by about 0.01.

[1] Data derived from the amino acid composition of foods [52]. The values for $EeqO_2$ are calculated from the metabolizable energy values of proteins (heat released on combustion, with aqueous urea as nitrogenous end product). The heat released during hydrolysis was assumed to be 20 kJ/mol peptide bond (fig. 1).

easily appreciated by considering the principles and methods involved in the calculations.

Oxidation of a Single Substrate

When only one substrate is oxidized to CO_2 and water, energy expenditure can be accurately predicted from a single measured parameter, e.g. O_2 consumption or CO_2 production multiplied by an appropriate calorimetric coefficient. Note that the energy released per litre O_2 consumed ($EeqO_2$) or per litre CO_2 produced ($EeqCO_2 = EeqO_2/RQ$) varies with the fuel (table 2).

Oxidation of a Two-Substrate Mixture

When two substrates are oxidized simultaneously (e.g. carbohydrate and fat) two measurement parameters are necessary to predict energy expenditure. The most commonly used parameters are O_2 consumption and CO_2 production. For an oxidation mixture of two fuels the ratio of CO_2 produced to O_2 consumed (the respiratory quotient; RQ) indicates the proportion of energy derived from each substrate. When the RQ is 1.0 (the RQ of carbohydrate) all the energy is derived from carbohydrate oxidation. When the RQ is 0.71 (the RQ of fat) all the energy is derived from fat oxidation. An intermediate RQ implies that both fat and carbohydrate contribute to energy expenditure. These considerations allow the derivation of formulae for calculating energy expenditure from measurements of O_2 consumption and CO_2 production. If the oxidation of a mixture of carbohydrate and fat consumes 1 litre O_2 and the proportion used in oxidizing carbohydrate is c, then the proportion used in fat oxidation is 1-c. The energy equivalent of 1 litre O_2 ($EeqO_2$) is given by equation 2:

$$EeqO_{2\,mix} = c \cdot EeqO_{2\,carb} + (1 - c)EeqO_{2\,fat} = c(EeqO_{2\,carb} - EeqO_{2\,fat}) + EeqO_{2\,fat} \quad (2)$$

The respiratory quotient of the mixture (RQ_{mix}) is given by equation 3:

$$RQ_{mix} = \frac{c \cdot RQ_{carb} + (1 - c)RQ_{fat}}{1} \quad (3)$$

where $c \cdot RQ_{carb}$ is the CO_2 produced from carbohydrate oxidation and RQ_{fat} (1-c) is the CO_2 produced from fat oxidation. The denominator of 1 represents the total O_2 consumption, 1 litre.

Collecting the terms for c in equation 3 gives equation 4:

$$c = \frac{RQ_{mix} - RQ_{fat}}{RQ_{carb} - RQ_{fat}} \quad (4)$$

Substituting c into equation 2 gives a solution to the $EeqO_2$ for the oxidation mixture (equation 5):

$$EeqO_{2\,mix} = \frac{(RQ_{mix} - RQ_{fat})(EeqO_{2\,carb} - EeqO_{2\,fat})}{RQ_{carb} - RQ_{fat}} + EeqO_{2\,fat} \quad (5)$$

Substituting the standard values of Eeq_{carb} Eeq_{fat}, RQ_{carb} and RQ_{fat} tabulated in table 1 gives equation 6:

$$EeqO_{2\ mix}\ (kJ/l) = 15.913 + 5.207\ RQ_{mix} \tag{6}$$

Total energy expenditure is the product of $EeqO_{2\ mix}$ (kJ/l) and the amount of O_2 consumed (litres) (equation 7):

$$EE\ of\ mixture\ (kJ) = 15.913\ O_2 + 5.207\ CO_2 \tag{7}$$

where O_2 and CO_2 represent the number of litres of O_2 consumed and produced respectively. It is obvious that different equations will result when different values of RQ_{fat}, RQ_{carb}, $EeqO_{2\ carb}$ and $EeqO_{2\ fat}$ are inserted into equation 5. For example, if the values for glucose (RQ, 1.0; $EeqO_2$, 20.84 kJ/l) are used in equation 5 instead of the values for polysaccharide (starch, glycogen), the $EeqO_2$ of the fat-carbohydrate fuel mixture is given by the following equation:

$$EE\ of\ mixture\ (kJ) = 16.599\ O_2 + 4.241\ CO_2 \tag{8}$$

The same approach can be used to show how the formulae for calculating the contribution of carbohydrate and fat to total energy expenditure in a two-component mixture is sensitive to the calorimetric coefficients for both carbohydrate and fat. The percentage of energy derived from carbohydrate oxidation ($E\%_{carb}$) is given by equation 9:

$$E\%_{carb} = \frac{100 \cdot energy\ from\ carbohydrate}{energy\ from\ carbohydrate + fat} = \frac{100c \cdot EeqO_{2\ carb}}{c \cdot EeqO_{2\ carb} + EeqO_{2\ fat}(1 - c)} \tag{9}$$

In equation 9, $c \cdot Eeq_{2\ carb}$ is the energy derived from carbohydrate oxidation and $EeqO_{2\ fat}\ (1 - c)$ is the energy derived from fat oxidation.

Substituting c from equation 4 into equation 9 gives a solution for $E\%_{carb}$ (equation 10):

$$E\%_{carb} = \frac{100\ EeqO_{2\ carb}\ (RQ_{mix} - RQ_{fat})}{EeqO_{2\ carb}\ (RQ_{mix} - RQ_{fat}) + EeqO_{2\ fat}\ (RQ_{carb} - RQ_{mix})} \tag{10}$$

By inserting the standard values of $EeqO_{2\ carb}$, $EeqO_{2\ fat}$, RQ_{fat} and RQ_{carb} (table 2) into equation 10, the following equation is produced (equation 11):

$$E\%_{carb} = \frac{2112\ (RQ_{mix} - 0.71)}{21.12\ (RQ_{mix} - 0.71) + 19.61\ (1 - RQ_{mix})} \tag{11}$$

Equations 10 and 11 show that estimates of $E\%_{carb}$ are sensitive to the calorimetric coefficients for carbohydrate and fat, and that when these coefficients are defined, $E\%_{carb}$ is dependent only on the RQ_{mix}.

Table 13. Heat (energy) equivalents of O_2 and CO_2, and an analysis of the oxidation of a fat-carbohydrate mixture (RQ < 1.0)[1]

RQ	Heat released from oxidation of carbo-hydrate[2], %	Heat released from oxidation of fat, %	Energy-gas equivalents				Substrate balances[3]			
			O_2		CO_2		carbohydrate per unit heat released		fat per unit heat released	
			kJ/l	(kcal/l)	kJ/l	(kcal/l)	g/MJ	(g/Mcal)	g/MJ	(g/Mcal)
0.71	0.00	100.00	19.610	(4.687)	27.620	(6.601)	0.00	(0.0)	−25.38	(−106.2)
0.72	3.70	96.30	19.662	(4.699)	27.308	(6.527)	−2.11	(−8.8)	−24.44	(−102.3)
0.73	7.39	92.61	19.714	(4.712)	27.006	(6.455)	−4.22	(−17.6)	−23.51	(−98.3)
0.74	11.05	88.95	19.766	(4.724)	26.711	(6.384)	−6.31	(−26.4)	−22.58	(−94.5)
0.75	14.70	85.30	19.818	(4.737)	26.424	(6.316)	−8.39	(−35.1)	−21.65	(−90.6)
0.76	18.33	81.67	19.870	(4.749)	26.145	(6.249)	−10.46	(−43.8)	−20.73	(−86.7)
0.77	21.93	78.07	19.922	(4.762)	25.873	(6.184)	−12.54	(−52.4)	−19.81	(−82.9)
0.78	25.52	74.48	19.974	(4.774)	25.608	(6.121)	−14.57	(−61.0)	−18.90	(−79.1)
0.79	29.09	70.91	20.027	(4.786)	25.350	(6.059)	−16.61	(−69.5)	−18.00	(−75.3)
0.80	32.64	67.36	20.079	(4.799)	25.098	(5.999)	−18.64	(−78.0)	−17.10	(−71.5)
0.81	36.18	63.82	20.131	(4.811)	24.853	(5.940)	−20.65	(−86.4)	−16.20	(−67.8)
0.82	39.69	60.31	20.183	(4.824)	24.613	(5.883)	−22.66	(−94.8)	−15.31	(−64.0)
0.83	43.19	56.81	20.235	(4.836)	24.379	(5.827)	−24.65	(−103.2)	−14.42	(−60.3)
0.84	46.67	53.33	20.287	(4.849)	24.151	(5.772)	−26.64	(−111.5)	−13.54	(−56.6)
0.85	50.13	49.87	20.339	(4.861)	23.928	(5.719)	−28.62	(−119.7)	−12.66	(−53.0)
0.86	53.57	46.43	20.391	(4.874)	23.710	(5.667)	−30.58	(−128.0)	−11.78	(−49.3)
0.87	57.00	43.00	20.443	(4.886)	23.498	(5.616)	−32.54	(−136.1)	−10.91	(−45.7)
0.88	60.41	39.59	20.495	(4.898)	23.290	(5.566)	−34.48	(−144.3)	−10.05	(−42.0)
0.89	63.80	36.20	20.547	(4.911)	23.087	(5.518)	−36.42	(−152.4)	−9.19	(−38.4)
0.90	67.17	32.83	20.599	(4.923)	22.888	(5.470)	−38.35	(−160.4)	−8.33	(−34.9)
0.91	70.53	29.47	20.651	(4.936)	22.694	(5.424)	−40.26	(−168.5)	−7.48	(−31.3)
0.92	73.87	26.13	20.703	(4.948)	22.504	(5.379)	−42.17	(−176.4)	−6.63	(−27.7)
0.93	77.19	22.81	20.756	(4.961)	22.318	(5.334)	−44.07	(−184.4)	−5.79	(−24.2)
0.94	80.50	19.50	20.808	(4.973)	22.136	(5.291)	−45.95	(−192.3)	−4.95	(−20.7)
0.95	83.79	16.21	20.860	(4.986)	21.958	(5.248)	−47.83	(−200.1)	−4.11	(−17.2)
0.96	87.07	12.93	20.912	(4.998)	21.783	(5.206)	−49.70	(−208.0)	−3.28	(−13.7)
0.97	90.32	9.68	20.964	(5.010)	21.612	(5.165)	−51.56	(−215.7)	−2.46	(−10.3)
0.98	93.56	6.44	21.016	(5.023)	21.445	(5.125)	−53.41	(−223.5)	−1.63	(−6.8)
0.99	96.79	3.21	21.068	(5.035)	21.281	(5.086)	−55.25	(−231.2)	−0.81	(−3.4)
1.00	100.00	0.00	21.120	(5.048)	21.120	(5.048)	−57.09	(−238.8)	0.00	(0.0)

[1] This table was constructed using equations 6 and 11, and the gross energy density of fat (39.4 kJ/g) and carbohydrate as glucose polymer (glycogen/starch) (17.517 kJ/g) respectively.
[2] The carbohydrate as polymer; $EeqO_2$ = 21.12 kJ/l; RQ = 1.00. When the glucose is the carbohydrate ($EeqO_2$ = 20.843 kJ/l; RQ = 1.00) the values in the table change a little, except those for carbohydrate balance, which will be 11.1 % higher at RQ 0.72 and 12.6 % higher at RQ 1.0. This is largely due to the lower energy density of glucose compared with starch/glycogen.
[3] Negative values indicate net oxidation, or utilization. The substrate 'balances', which are calculated from gaseous exchange, do not include polysaccharide-glucose interconversions, which are not associated with gaseous exchange.

Table 14. Heat (energy) equivalents of O_2 and CO_2, and an analysis of carbohydrate utilized and fat accumulated ($RQ > 1.0$)[1]

| RQ | % carbohydrate energy utilized | | Energy-gas equivalents | | | | Substrate balances[3] | | | |
| | released as heat % | stored as fat[2] % | O_2 | | CO_2 | | carbohydrate per unit heat released | | fat per unit heat released | |
			kJ/CO_2	(kcal/CO_2)	kJ/l CO_2	(kcal/l)	g/MJ	(g/Mcal)	g/MJ	(g/Mcal)
1.00	100.00	0.00	21.120	(5.048)	21.120	(5.048)	−57.09	(−238.8)	0.00	(0.0)
1.01	96.91	3.09	21.172	(5.060)	20.962	(5.010)	−58.91	(−246.5)	0.81	(3.4)
1.02	94.01	5.99	21.224	(5.073)	20.808	(4.973)	−60.72	(−254.1)	1.62	(6.8)
1.03	91.30	8.70	21.276	(5.085)	20.657	(4.937)	−62.53	(−261.6)	2.42	(10.1)
1.04	88.75	11.25	21.328	(5.098)	20.508	(4.902)	−64.33	(−269.1)	3.22	(13.5)
1.05	86.35	13.65	21.380	(5.110)	20.362	(4.867)	−66.11	(−276.6)	4.01	(16.8)
1.06	84.08	15.92	21.432	(5.122)	20.219	(4.833)	−67.89	(−284.1)	4.80	(20.1)
1.07	81.95	18.05	21.484	(5.135)	20.079	(4.799)	−69.66	(−291.5)	5.59	(23.4)
1.08	79.02	20.08	21.537	(5.147)	19.941	(4.766)	−71.42	(−298.8)	6.38	(26.7)
1.09	78.01	21.99	21.589	(5.160)	19.806	(4.734)	−73.18	(−306.2)	7.15	(29.9)
1.10	76.19	23.81	21.641	(5.172)	19.673	(4.702)	−74.92	(−313.5)	7.93	(33.2)
1.11	74.47	25.53	21.693	(5.185)	19.543	(4.671)	−76.66	(−320.7)	8.70	(36.4)
1.12	72.82	27.18	21.745	(5.197)	19.415	(4.640)	−78.39	(−328.0)	9.47	(39.6)
1.13	71.50	28.50	21.797	(5.210)	19.289	(4.610)	−80.11	(−335.2)	10.24	(42.8)
1.14	69.77	30.23	21.849	(5.222)	19.166	(4.581)	−81.82	(−342.3)	11.00	(46.0)
1.15	68.35	31.65	21.901	(5.234)	19.044	(4.552)	−83.52	(−349.5)	11.75	(49.2)
1.16	66.99	33.01	21.953	(5.247)	18.925	(4.523)	−85.22	(−356.6)	12.51	(52.3)
1.17	65.69	34.31	22.005	(5.259)	18.808	(4.495)	−86.91	(−363.6)	13.26	(55.5)
1.18	64.44	35.56	22.057	(5.272)	18.693	(4.468)	−88.59	(−370.6)	14.01	(58.6)
1.19	63.25	36.75	22.109	(5.284)	18.579	(4.441)	−90.26	(−377.6)	14.75	(61.7)
1.20	62.10	37.90	22.161	(5.297)	18.468	(4.414)	−91.92	(−384.6)	15.49	(64.8)
1.21	61.00	39.00	22.213	(5.309)	18.358	(4.388)	−93.58	(−391.5)	16.23	(67.9)
1.22	59.95	40.05	22.266	(5.322)	18.250	(4.362)	−95.23	(−398.4)	16.96	(71.0)
1.23	58.93	41.07	22.318	(5.334)	18.144	(4.337)	−96.87	(−405.3)	17.69	(74.0)
1.24	57.95	42.05	22.370	(5.346)	18.040	(4.312)	−98.50	(−412.1)	18.41	(77.0)
1.25	57.01	42.99	22.422	(5.359)	17.937	(4.287)	−100.13	(−418.9)	19.14	(80.1)
1.26	56.11	43.89	22.474	(5.371)	17.836	(4.263)	−101.74	(−425.7)	19.86	(83.1)
1.27	55.23	44.77	22.526	(5.384)	17.737	(4.239)	−103.35	(−432.4)	20.57	(86.1)
1.28	54.39	45.61	22.578	(5.396)	17.639	(4.216)	−104.96	(−439.1)	21.28	(89.1)
1.29	53.58	46.42	22.630	(5.409)	17.543	(4.193)	−106.55	(−445.8)	21.99	(92.0)
1.30	52.79	47.21	22.682	(5.421)	17.448	(4.170)	−108.14	(−452.5)	22.70	(95.0)

[1] This table was constructed using equations 6 and 11 and the gross energy density of fat (39.4 kJ/g) and carbohydrate as glucose polymer (glycogen/starch) (17.517 kJ/g).

[2] The carbohydrate polymer; $EeqCO_2 = 21.12$ kJ/l; RQ 1.00. If glucose is the carbohydrate ($EeqO_2 = 20.843$ kJ/l; RQ = 1.00) the values in the table will change little, except those for carbohydrate balance, which will be 12.6 % higher at RQ 1.00 and 13.3 % higher at RQ 1.30. This is largely due to the lower energy density of glucose compared to starch/glycogen (glucan).

[3] Negative values indicate net utilization and positive values net synthesis. The substrate 'balances', which are calculated from gaseous exchange, do not include glucan-glucose interconversions, which are not associated with gaseous exchange.

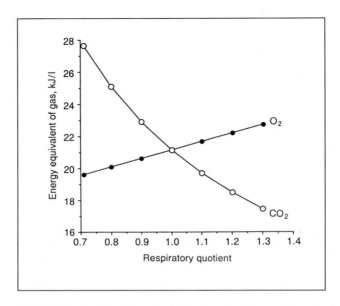

Fig. 3. Relationship between respiratory quotient and the energy equivalents of O_2 (●) and CO_2 (○) for a fat-carbohydrate mixture. Below a respiratory quotient of 1.00 there is oxidation of fat and carbohydrate. Above a respiratory quotient of 1.00 there is carbohydrate oxidation and net conversion of carbohydrate to fat.

Table 13 shows how the percentage of energy derived from carbohydrate and fat oxidation respectively, and $EeqO_2$ $_{mix}$ and $EeqCO_2$ $_{mix}$ vary with RQ_{mix} for a fat-carbohydrate oxidation mixture. The table also indicates the carbohydrate and fat utilized per unit heat produced. Table 14 shows similar parameters for values of $RQ_{mix} > 1$ which occur when there is net fat synthesis from carbohydrate. The validity of calorimetry during net lipid synthesis has been discussed in considerable detail elsewhere [41], and is briefly considered again later in this article.

Essentially, the equations used to calculate energy expenditure and fuel selection during net fat oxidation can also be used during net fat synthesis from carbohydrate (see below). Numerically, lipogenesis from glucose is equal to 'negative fat oxidation'. The $EeqO_2$ $_{mix}$ and the $EeqCO_2$ $_{mix}$ each show a smooth transition above and below the RQ_{mix} of 1.0 (fig. 3). Similarly, $E\%_{carb}$ and $E\%_{fat}$ follow a smooth transition above and below the RQ of 1.0 (fig. 4).

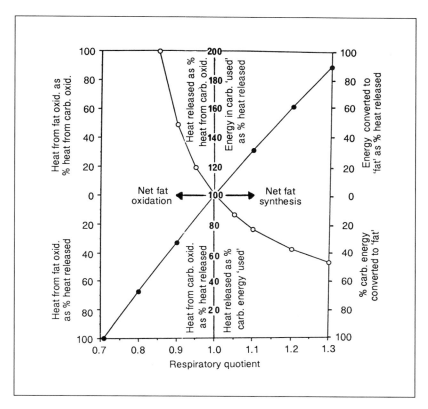

Fig. 4. Dependence of carbohydrate utilization and fat oxidation and synthesis on the respiratory quotient (RQ_{mix}). The two curves describe carbohydrate (starch or glycogen) utilization and fat oxidation and synthesis, with the ordinate axis partitioned into several variables. 'Carbohydrate used' refers to the carbohydrate involved in gaseous exchange. It does not include interconventions of glucose monosaccharide to glucose polysaccharides and vice versa [based on 41].

Oxidation of a Three-Substrate Mixture

When three substrates are oxidized simultaneously it is necessary to have three measurements in order to calculate energy expenditure and fuel selection. In the subsequent example, protein is chosen as the third substrate, because it is commonly involved in biological systems. The two most common methods used for assessing the rate of protein oxidation are the rate of urinary N excretion and the rate at which carbon-labelled CO_2 is expired after infusion of carbon-labelled amino acids. Both methods have problems. With the labelling techniques assumptions have to be made about equilibration of the labelled amino acid within the body pool, and about the contribution of that amino acid to total amino acid oxidation. Furthermore, assumptions have to be made about the extent of recovery of the labelled CO_2. This is because when labelled $^{14}CO_2$ or $^{13}CO_2$ is infused into subjects in the form of bicarbonate, the recovery of the label in breath is incomplete due to entry of the label into pools of CO_2 that turn over slowly, e.g. in bone, and to fixation of labelled carbon in fat, carbohydrate and protein [3].

With the urinary N method, errors may arise if urine collections are incomplete or if urine is not completely voided from the bladder between the beginning and end of the study. Further difficulties may arise if the end products of metabolism, such as urea, accumulate in the blood. For example, when the blood urea concentration rises by 1 mmol/l (about 20%), in a 70-kg man, this is associated with the accumulation of about 42 mmol urea (1.18 gN) in the body pool. This is because the distribution volume of urea, in litres, is about $0.6 \times$ body weight (kg). However, because the rate of N excretion is most frequently used as an indicator of protein oxidation in studies of indirect calorimetry, it is used in the example below for deriving the calorimetry equations for three substrate mixtures.

If a mixture of carbohydrate, protein and fat consumes 1 litre O_2, and the proportions used in the oxidation of carbohydrate, protein and fat are c, p and f respectively (equation 12) (as used by Weir [18]):

$$1 = c + p + f \tag{12}$$

Since the respiratory quotients of carbohydrate, protein and fat are 1.0, 0.835 and 0.71 respectively (table 2), an equation can also be written for RQ_{mix} (equation 13):

$$RQ_{mix} = c + 0.835p + 0.71f \tag{13}$$

RQ_{mix} is the combined amount of CO_2 produced from the oxidation of carbohydrate (c), protein (0.835p) and fat (0.71f) – per litre of O_2 consumed. The energy expenditure associated with the consumption of 1 litre O_2 (i.e. $EeqO_2$) is given by equation 14:

$$EeqO_{2\ mix}\ (kJ/l) = 21.12c + 19.48p + 19.61f \tag{14}$$

where 21.12, 19.48 and 19.61 kJ/l are the $EeqO_2$'s for carbohydrate, protein and fat respectively (see table 2).

Substituting c from equation 12 into equations 13 and 14 give equations 15 and 16 respectively:

$$RQ_{mix} = 1 - p - f + 0.835p + 0.71f \tag{15}$$

$$EeqO_{2\ mix} = 21.12 - 1.64p - 1.51f \tag{16}$$

Substituting f from equation 15 into equation 16 gives equation 17:

$$EeqO_{2\ mix} = 15.913 + 5.207\ RQ_{mix} - 0.7809p \tag{17}$$

The first two terms are identical to those in equation 7 for a fat-carbohydrate mixture and the last term is a modifying term due to protein oxidation.

The term p, the proportion of O_2 consumption due to the oxidation of protein, can be transformed to the measured quantity of urinary nitrogen, Nu. Since 5.95 litres O_2 are consumed for each g of Nu, then 0.7809p is equal to 4.646 Nu(g), and equation 17 becomes equation 18:

$$EeqO_{2\ mix}\ (kJ/l\ O_2) = 15.913 + 5.207\ RQ_{mix} - 4.646\ Nu(g) \tag{18}$$

where Nu(g) is the urine N excreted during the time it takes to use 1 litre O_2.

Energy expenditure ($EeqO_{2\ mix}\cdot$total O_2 consumed) is given by equation 19:

$$EE\ (kJ) = 15.913\ O_2 + 5.207\ CO_2 - 4.646\ Nu(g) \tag{19}$$

where Nu(g) is the urine N excreted during the measurement period.

The first two terms of equation 19 are identical to those in equation 7 for a fat-carbohydrate mixture. The term 4.646 Nu(g) can be regarded as a modification to equation 7 to account for the presence of protein in the oxidation mixture.

Oxidation of a Four (or More)-Substrate Mixture

For each additional substrate in an oxidation mixture, an additional measurement needs to be made and an additional modifying term needs to be inserted in the prediction equation for $EeqO_2$. For example, when alcohol is part of a mixture with fat, protein and carbohydrate, the proportion of O_2 consumed during the oxidation of carbohydrate (c), protein (p), fat (f) and alcohol (a) are indicated by equation 20:

$$1 = c + p + f + a \tag{20}$$

The RQ_{mix} is indicated by equation 21, and $EeqO_{2mix}$ by equation 22:

$$RQ_{mix} = c\cdot RQ_{carb} + p\cdot RQ_{prot} + f\cdot RQ_{fat} + a\cdot RQ_{alc} \tag{21}$$

$$EeqO_{2\ mix} = c\cdot EeqO_{2\ carb} + p\cdot EeqO_{2\ prot} + f\cdot EeqO_{2\ fat} + a\cdot EeqO_{2\ alc} \tag{22}$$

Substitution of c and f from equations 20 and 21 respectively and insertion of values for $EeqO_2$ for the different fuels from table 2 gives equation 23:

$$EeqO_{2\,mix} \text{ (kJ/l } O_2) = 15.913 + 5.207\ RQ - 0.781p + 0.944a \tag{23}$$

Energy expenditure ($EeqO_{2\,mix} \cdot$ total O_2 consumed) is given by equation 24:

$$EE \text{ (kJ)} = 15.913\ O_2 + 5.207\ CO_2 - 0.781\ pO_2 + 0.944\ aO_2 \tag{24}$$

Since 1 gNu is equivalent to the utilization of 5.95 litres of O_2 ($pO_2 = 5.95 \times$ gNu) and 1 g alcohol requires 1.460 litres of O_2 for complete oxidation ($aO_2 = 1.460 \times$ g alcohol), equation 24 becomes equation 25:

$$EE \text{ (kJ)} = 15.913\ O_2 \text{ (l)} + 5.207\ CO_2 \text{ (l)} - 4.646\ Nu(g) + 1.380\ \text{alcohol (g)} \tag{25}$$

The first two terms in equation 25 are identical to those in equation 7 which applies to the simple, two-component, carbohydrate-fat oxidation mixture. The first three terms are identical to those in equation 19, which applies to the three-component carbohydrate-fat-protein oxidation mixture. The fourth term in equation 25 is the additional modifying term which is necessary for calculating energy expenditure when alcohol is a component of the oxidation mixture.

In man virtually all ingested alcohol is absorbed and oxidized. The amount ingested therefore provides a good indication of the amount oxidized – provided that a sufficiently long time has been allowed for absorption and oxidation to be complete.

Partial Oxidation of Substrates in Mixtures

The procedure used to establish the equations for a carbohydrate-fat-protein oxidation mixture takes into account the partial oxidation of protein to combustible nitrogenous end products of metabolism (e.g. urea, ammonia). The same procedure can also be used to establish equations that apply to the partial oxidation of other substrates such as fat or carbohydrate. However, the quantities of substrates metabolized and end products formed must be known.

If in a three-component mixture (carbohydrate-fat-protein) some of the carbohydrate is converted to H_2 as a result of fermentation, then the proportion of O_2 used in carbohydrate oxidation is c-h, where c is the proportion of O_2 that is necessary to completely oxidize the carbohydrate utilized, and h is the proportion of total oxygen that is required to oxidize the hydrogen formed (i.e., h represents the proportion of total O_2 consumption that has been spared as a result of the incomplete oxidation of carbohydrate). From this, the following three equations emerge:

$$1 = c + p + f - h \tag{26}$$

$$RQ_{mix} = c \cdot RQ_{carb} + f \cdot RQ_{fat} + p \cdot RQ_{prot} - h \cdot RQ_{hydr} \tag{27}$$

$$EeqO_{2\,mix} \ (kJ/l\ O_2) = c \cdot EeqO_{2\,carb} + f \cdot EeqO_{2\,fat} + p \cdot EeqO_{2\,prot} - h \cdot EeqO_{2\,hyd} \tag{28}$$

Substituting c and f from equations 26 and 27 respectively into equation 28 gives equations 29 and 30:

$$EeqO_{2\,mix} \ (kJ/l\ O_2) = 15.913 + 5.207\ RQ - 0.781p - 9.587h \tag{29}$$

$$EE_{total} \ (kJ) = 15.913\ O_2 + 5.207\ CO_2 - 0.781\ pO_2 - 9.587\ hO_2 \tag{30}$$

Again, 1 gNu is equivalent to the utilization of 5.95 litres O_2 (pO_2 = 5.95 × gNu) and 1 litre of H_2 requires 0.5 litre O_2 for complete oxidation (hO_2 = 0.5 × litres H_2), and equation 30 becomes equation 31:

$$EE\ (kJ) = 15.913\ O_2 + 5.207\ CO_2 - 4.646\ gNu - 4.793\ H_2 \tag{31}$$

Again the first three terms are identical to those in equation 19. The overall estimate of energy expenditure requires knowledge of the amount of hydrogen produced (the fourth measurement for this four-component mixture). Similar calculations can be made to establish modifying or adjustment terms for other end products of metabolism. The adjustment terms are constants provided the RQ and $EeqO_2$ associated with the complete metabolic oxidation of the other components are also constant. A simple method for deriving adjustment terms is considered below.

A Simple Method to Modify a General Equation for Calculating Energy Expenditure when Additional Substrates Are Used and Additional End Products Are Formed

The derivation of a simple method to obtain modifying terms can be illustrated using the example of a three-component mixture in which 1 litre of O_2 is associated with the utilization of f litres O_2 for complete oxidation of fat, c litres O_2 for the complete oxidation of carbohydrate and z litres O_2 for complete metabolic oxidation of a third substance Z, the following relationship applies (equation 32):

$$1 = c + f + z \tag{32}$$

If substance Z is an end product of carbohydrate or fat metabolism, z has a negative sign.

Following the same procedures as before, equations 33 and 34 are established:

$$RQ_{mix} = c \cdot RQ_{carb} + f \cdot RQ_{fat} + z \cdot RQ_{subst \cdot z} \tag{33}$$

$$EeqO_{2\ mix} = c \cdot EeqO_{2\ carb} + f \cdot EeqO_{2\ fat} + z \cdot EeqO_{2\ subst \cdot z} \tag{34}$$

Substituting c from equation 32 and f from equation 33 respectively into equation 34 gives the following equation:

$$EeqO_{2\ mix} = 15.913 + 5.207\ RQ_{mix} - 5.207z(RQ_{subst \cdot z} - 0.71) + \tag{35}$$
$$z(EeqO_{2\ subst \cdot z} - 19.61)$$

Multiplying by O_2 consumption in litres gives equation 36:

$$EE = 15.913\ O_2 + 5.207\ CO_2 - 5.207\ zO_2(RQ_{subst \cdot z} - 0.71) + \tag{36}$$
$$zO_2\ (Eeq_{subst \cdot z} - 19.61)$$

Again, the first two terms of equation 36 appeared in equations 6, 7, 18, 23 and 31 for the two or more component mixtures that include the same carbohydrate and fat.

A solution to the remaining terms gives the modifying term for the substance, $K_{subst \cdot z}$ such that:

$$EE = 15.913\ O_2 + 5.207\ CO_2 + K_{subst \cdot z} \tag{37}$$

where

$$K_{subst \cdot z} = zO_2[-5.207(RQ_{subst \cdot z} - 0.71) + (EeqO_{2\ subst \cdot z} - 19.61)] \tag{38}$$

The term zO_2 is the amount of oxygen used in the oxidation of alcohol. Table 15 includes values for RQ, $EeqO_2$, zO_2 for various substances and the calculated values of $K_{subst \cdot z}$.

Effects of Net Lipogenesis from Carbohydrate on the Equations for Predicting Energy Expenditure and Fuel Selection

There has been controversy [see 41] as to whether the equation relating energy expenditure to the oxidation of carbohydrate and fat (equation 7) also applies to situations where there is net fat synthesis from carbohydrate. However, if in equation 38, substance Z is newly formed fat with an RQ of 0.71 and $EeqO_{2\ fat}$ of 19.61 kJ/l, then the calculated value for $K_{subst \cdot z}$ (equation 38) becomes zero and equation 37 reduces to equation 7. Consequently, equation 7 and its immediate predecessor, equation 6, which apply to situations where there is net fat oxidation that does not have to be adjusted before application to situations where there is net fat synthesis. This mathematical procedure provides another formal proof [see 41] of the validity of indirect calorimetry during net lipid synthesis. It can also be shown that the equations used to partition the energy between different fuels (e.g. equation 11) also apply during net lipid synthesis from carbohydrate [41]. Although $E\%_{carb}$ in equations 7 and 11 refers to % energy derived from carbohydrate oxidation when the RQ is ≤ 1.0, when there is net lipid synthesis from carbohydrate and when the RQ > 1.0, $E\%_{carb}$ refers to the net carbohydrate 'used' or 'utilized' since some of it is converted to fat (i.e. that carbohydrate involved in gaseous exchange).

A General Equation for Calculating Energy Expenditure

Finally, the constants tabulated in table 15 can be used to construct a general formula for calculating energy expenditure (equation 39):

$$
\begin{aligned}
EE \text{ (kJ)} = \; &15.913 \; O_2 \text{ (l)} \\
&+5.207 \; CO_2 \text{ (l)} \\
&-4.646 \; Nu \text{(g)} \\
&+1.380 \; \text{Alcohol oxidized (g)} \\
&+0.677 \; \text{Free glycerol oxidized (g)} \\
&-4.793 \; H_2 \text{ produced (l)} \\
&-2.686 \; CH_4 \text{ produced (l)} \\
&+0.632 \; \text{Hydroxybutyrate produced (g)} \\
&+1.159 \; \text{Acetoacetate produced (g)} \\
&-0.234 \; \text{Acetone produced (g)}
\end{aligned}
\tag{39}
$$

where fat, carbohydrate, glycerol and protein are oxidized to yield water, CO_2, nitrogenous end products (urea, ammonia and creatinine in a N mass ratio of 95:5:5, respectively) methane hydrogen and ketone bodies (acetoacetate, hydroxybutyrate, acetone) which are excreted or accumulate in the body. If there is no net oxidation of alcohol and free glycerol (distinct from glycerol in triglyceride which is taken into account in the basic fat-carbohydrate equation) and no net accumulation or excretion of ketone bodies, the equation simplifies to the first three terms which involve fat and carbohydrate (first two terms) and protein (third term).

When the rate of protein oxidation is not known, reasonable estimates of energy expenditure can still be made by making approximate assumptions about the rate of protein oxidation. The following procedure, which was also used by Weir [18], illustrates the extent of errors arising from such assumptions. If the proportion of total energy expenditure derived from protein oxidation is $Eprop_{prot}$, then the proportion of O_2 used in protein oxidation is $Eprop_{prot} \cdot EeqO_{2 \, mix}/EeqO_{2 \, prot}$. This term can be substituted for the term p from equation 17 ($EeqO_{2 \, mix} = 15.913 + 5.207RQ_{mix} - 0.7809p$) to produce equation 40:

$$EeqO_{2 \text{ mix}} = 15.913 + 5.207 \, RQ_{\text{mix}} - \frac{0.7809 \, EeqO_{2 \text{ mix}} \cdot Eprop_{\text{prot}}}{19.48} \qquad (40)$$

where 19.48 (kJ/l) is the $EeqO_{2 \text{ prot}}$ (table 2).

On rearranging and multiplying by O_2 consumption, equation 40 becomes equation 41:

$$EE \, (kJ) = \frac{15.913 \, O_2 + 5.207 \, CO_2}{1 + 0.040 \, Eprop_{\text{prot}}} \qquad (41)$$

When energy expenditure is calculated without the protein correction factor (i.e. using only the numerator in equation 41 (this is identical to equation 7), there is 0.4% error in energy expenditure for each 10% of total energy expenditure that arises from protein oxidation. When $Eprop_{\text{prot}} = 0.15$ (i.e. when protein oxidation accounts for 15% energy expenditure), equation 41 simplifies to equation 42:

Table 15. Values used in equation 37 for the calculation of the adjustment factors (K)

Substance	Molec-ular weight	Heat of combustion		EeqO$_2$ of substrate
		kJ/g or kJ/l	kJ/mol	kJ/l O$_2$
Protein[3] (utilization with excretion of urinary N (Nu))				19.48
Alcohol (oxidation)	46.07	29.68 kJ/g	−1,367	20.33
Glycerol (oxidation)	92.11	−18.03 kJ/g	−1,661	21.17
Methane (production)	16.04	−39.71 kJ/l	−890	19.86
Hydrogen (production)	2.016	−12.75 kJ/l	−286	25.50
3-Hydroxybutyric acid (BOH) (production)	104.1	−19.27 kJ/g	−2,006	19.89
Acetoacetic acid (AcAc) (production)	102.9	−17.39 kJ/g	−1,775	19.80
Acetone (AC) (production)	58.08	−30.83 kJ/g	−1,790	19.97

[1] The factor zO_2 in equation 37 is the product of the tabulated values (litres O_2 used per g (or l or mol) and the amount (g, l, or mol) of substrate oxidized. The +ve signs indicate that the substance is oxidized and the −ve signs that the substance is produced (equivalent to negative oxidation).

[2] The sign of the adjustment factor changes depending on whether the substance is oxidized or produced. The signs indicated here refer to the oxidation of protein alcohol and glycerol, and the production of the other substances (left column).

$$EE \text{ (kJ)} = 15.818 \, O_2 + 5.176 \, CO_2 \tag{42}$$

The modifying terms shown in table 15 for various substrates and products can also be used with equation 42 (except protein which is already considered in this equation).

Estimating the Proportion of Energy Derived from the Oxidation of Different Fuels

If a mixture of fat, protein and carbohydrate are oxidized simultaneously, it is possible to partition the energy expenditure from each source using three measurements: O_2 consumption; CO_2 production, and urine N excretion. The following three steps are necessary (table 16):

RQ of sub-stance	Litres O_2 used[1]		K (adjustment factor) kJ[2]	
	per g (or per l) oxidized	per mol oxidized	kJ/g (or kJ/l) $\times$ g (or l) oxidized	kJ/mol$\times$mol oxidized
0.835	+5.95/gNu	83.30/mol Nu	−4.646×gNu	−65.04×mol Nu
0.6667	+1.460/g	+67.24	+1.380×g alcohol	+63.58×mol alcohol
0.857	+0.852/g	+78.45	+0.677×g glycerol	+62.37×mol glycerol
0.500	−2.000/l	−44.83	−2.687×l CH_4	+60.22×mol CH_4
0.000	−0.500/l	−11.21	−4.793×l H_2	−107.44×mol H_2
0.889	−0.969/g	−100.86	+0.632×g BOH	+65.77×mol BOH
1.000	−0.878/g	−89.66	+1.159×g AcAc	−118.35×mol AcAc
0.750	−1.544/g	−89.66	−0.234×g AC	−13.60×mol AC

[3] Oxidized to urea, ammonia and creatinine in the nitrogenous ratio of 95:5:5. All the values apart from RQ refer to urinary N (Nu); (6.25 protein being equivalent to 1 gNu).

Table 16. Calculation of energy expenditure and fuel selection from gaseous exchange and urine N excretion (500 litres O_2, 425 litres CO_2 and 12 g urine N)

		O_2, l	CO_2, l	Energy expenditure kJ[1]
1	Total gaseous exchange	500	425	–
2	Protein oxidation (12 gNu)[2]	71.4	59.6	1,392
3	Alcohol oxidation	–	–	–
4	Fat and carbohydrate oxidation[3]	428.6	365.4	8,723
5	Carbohydrate oxidation[4]	210.7	210.7	4,449
6	Fat oxidation (row 4 minus row 5)	217.9	154.7	4,274
7	Total energy expenditure (row 2 plus row 4)	–	–	10,115

[1] The values for energy expenditure are essentially identical to those given in table 19 (last row), which were calculated using a different procedure (see text).
[2] O_2 consumed in protein oxidation = 12 gNu $\times$ O_2:Nu = 71.4 litres (12 $\times$ 5.95) (see tables 2 and 15). CO_2 consumed in protein oxidation = O_2 $\times$ RQ_{prot} = 59.619 litres. Energy from protein oxidation = gNu $\times$ 116 = 1,392 kJ where 116 is the heat released per g urinary N (see tables 2 and 15).
[3] Energy expenditure from fat + carbohydrate calculated using equation 7.
[4] Energy expenditure from carbohydrate calculated using equation 11. O_2 consumption and CO_2 production calculated by dividing energy derived from carbohydrate by 21.12 kJ/l (the values EeqO$_2$ carb and EeqCO$_2$ carb; table 2).

(1) Using the measurement of urine N excretion, the energy expenditure, O_2 consumption, and CO_2 production associated with protein oxidation are determined. One gram of urinary N is considered to be equivalent to 116 kJ, 5.95 litres O_2 and 4.97 litres CO_2 (tables 2, 15) [41].

(2) Gaseous exchange due to the utilization of substrates in a fat and carbohydrate mixture is calculated by deducting from the total gaseous exchange, that attributable to protein oxidation (table 16).

(3) The energy derived from fat and carbohydrate oxidation mixture is calculated using equation 7 and the individual contributions of fat and carbohydrate are calculated using equation 11.

It is possible to alter the order in which some of the calculations are made without changing the results. For example, for a protein-fat-carbohydrate oxidation mixture, total energy expenditure can be calculated first using equations 19 or 39. Nonprotein energy expenditure and gaseous exchange can then be calculated by subtracting the energy expenditure and gaseous exchange attributable to protein oxidation from total energy expenditure and gaseous exchange, respectively.

If a fourth substrate is present in the oxidation mixture it is possible to analyse the contribution of each substrate to energy expenditure provided known quantities of the fourth substrate are metabolized to known quantities of end products. For example, if it is assumed that all ingested alcohol is oxidized to CO_2 and water (or more correctly 98% oxidized since some alcohol is excreted in urine and breath [42]), then the energy expenditure and gaseous exchange associated with alcohol metabolism can be calculated. The oxidation of 1 g alcohol is associated with the production of 29.68 kJ, consumption of 1.46 litres O_2, and release of 0.973 litres CO_2; $EeqO_2$ alc, 20.33 kJ/l; molecular weight, 46.07 daltons. The gaseous exchange and energy expenditure associated with alcohol and protein oxidation can be subtracted from total gaseous exchange, to provide a solution to the contribution of the fat-carbohydrate mixture to energy expenditure (equation 11).

Estimation of ATP Gain

It has been suggested [43] that ATP acts as a form of 'biochemical currency'. The amount of ATP gained might be a useful indicator of energy transformations and energy efficiency, and better than the amount of heat released [10, 34, 41]. Unfortunately, calculations of the amount of ATP gained is not an easy task because it depends on the biochemical pathways involved and on the degree of mitochondrial coupling of oxidative phosphorylation, both of which are somewhat uncertain [10, 44]. If the degree of coupling is assumed and the pathways defined, the energy equivalents of ATP can be calculated as the ratio of ATP gained to heat released (mol/kJ). Here it is important to make a distinction between 'ATP gained' and 'ATP produced'. In a biochemical pathway some of the ATP produced is used obligatorily. In the case of glucose oxidation, 1-ATP is used in the conversion of glucose to glucose-6-phosphate and another ATP for the conversion of fructose-6-phosphate to fructose-1,6-bisphosphate. Therefore, two extra moles of ATP are produced than are gained. The 'ATP gained' does not mean ATP accumulates since ATP turns over rapidly. Rather, it represents the ATP which is available to the body as a 'fuel' for other metabolic processes. Substrate cycles such as the Cori cycle and triglyceride-fatty acid cycle [41, 45], are examples of such processes which require ATP.

Table 17. The metabolizable energy equivalents of ATP

	Enthalpy change kJ/mol	ATP gained per mol substrate[5]	Energy equivalent of ATP gained kJ/mol ATP
Fat (dioleylpalmitate) (oxidation via β-oxidation and citric acid cycle)	34,022[1]	429.4	79.2
Glycogen [$(C_6H_{10}O_5)_n$] (oxidation via glycolysis and citric acid cycle)	2,840[2]	37.7[2]	75.3
Starch [$(C_6H_{10}O_5)_n$] (oxidation via glycolysis and citric acid cycle)	2,840[2]	36.7[2]	77.4
Glucose $C_6H_2O_6$ (oxidation via glycolysis and citric acid cycle)	2,803	36.7	76.4
Protein (direct oxidation)	–	–	86.9
Protein (oxidation via gluconeogenesis)	–	–	91.3
Glycerol (oxidation via glycolysis and citric acid cycle)	1,661	21.0[3]	79.1[3]
Glucose → fat (dioleylpalmitate):			
With maximum malate cycle	8,023[4]	26.4[4]	303.9
With maximum pentose phosphate pathway	14,077[4]	119.7[4]	117.6
Starch → fat (dioleylpalmitate):			
With maximum malate cycle	8,578[4]	26.4[4]	324.9
With maximum pentose phosphate pathway	14,712[4]	119.7	122.9

[1] The heat of combustion of dioleylpalmitate (RQ 0.71) was calculated from the sum of the heats of combustion of the constituent fatty acids and glycerol (see section on 'Establishing the Calorimetric Coefficients for Fat, Carbohydrate and Proteins' [cf. 41].

[2] Per n.

[3] Assumes that the conversion of glycerol-3-phosphate to dihydroxyacetone phosphate is linked to the formation of $NADH_2$ from NAD. When it is linked to the formation of $FADH_2$ from FAD the total ATP gain is 20.33 and the energy equivalent of ATP gained is 81.7 kJ/mol ATP.

[4] Per mol fat (dioleylpalmitate) formed.

[5] Calculated as the cytoplasmic ATP equivalent [10].

Table 17 indicates the values for ATP gained [41] and the energy equivalents of ATP, which are derived by considering classic biochemical pathways and a likely degree of coupling of oxidative phosphorylation [see 10, 41, 44]. During net fat synthesis more energy (kJ) has to be expended per ATP gained than during either net fat or carbohydrate oxidation. The energy equivalent of 'ATP gained' increases abruptly above a nonprotein RQ of 1.0 (fat-carbohydrate mixture). Therefore, in contrast to the standard energy equation which can be used to calculate energy expenditure (heat exchange) both above and below a nonprotein RQ of 1.0 (see 'Effects of Net Lipogenesis from Carbohydrate on the Equations for Predicting Energy Expenditure and Fuel Selection'), this cannot be done for ATP

Table 18. Calorimetric coefficients for carbohydrate, fat and protein suggested by various authors (see also tables 19 and 20)

Author	Carbohydrate		Fat		Protein			
	EeqO$_2$ kJ/l	RQ	EeqO$_2$ kJ/l	RQ	EeqO$_2$ kJ/l	RQ	O$_2$/Nu l/g	kJ/Nu l/g
Zuntz [25]	21.12	1.00	19.61	0.707	18.73	0.793	6.06	114
Loewy [16]/Lusk [15][1], Abramson [19]	21.12	1.00	19.61	0.707	18.67	0.8016	5.94	111
Magnus-Levy [17][2]	21.12	1.00	19.61	0.707	19.25	0.809	6.04[2]	116[2]
Erwin et al. [39]	21.09	–	19.57	–	19.66	–	–	–
Brody [40]	21.12	1.00	19.62	0.710	20.17	0.82	–	–
Weir [18][3]	21.12	1.00	19.81	0.718	18.68	0.802	5.94	111
Consolazio et al. [21], Passmore and Eastwood [22][4]	20.67	1.00	19.27	0.707	17.78	0.809	6.03	107
Benedict and Talbot [46], Carpenter [47], Brouwer [38][5]	21.19	1.00	19.74	0.711	19.24	0.809	5.98	115
Brockway [48]	21.10	1.00	19.81	0.715	19.25	0.809	5.98	115
Hunt [27][6] 'fasting'	20.69	1.00	19.27	0.703	17.80	0.820	–	–
'fed'	20.69	1.00	20.86	0.703	20.69	0.820	–	–
Bursztein et al. [28, 29][6] 'fasting'	20.69	1.00	19.27	0.707	17.76	0.809	6.04	107
'fed'	20.69	1.00	20.86	0.703	20.96	0.820	5.18	107
Ben-Porat et al. [23][7]	20.95	1.00	19.60	0.705	19.93	0.809	5.13[7]	102
Livesey and Elia [7], and present paper	21.12	1.00	19.61	0.710	19.48	0.835	5.95	116

[1] Values of Loewy [16] modified by Lusk [15] (see text).

[2] O$_2$:Nu and kJ/gN assume a protein:N conversion factor of 6.25 as suggested by Consolazio et al. [21], Ben-Porat et al. [23] and Bursztein et al. [28, 29] (see text).

[3] Carbohydrate factors from Zuntz [25], fat from Cathcart and Cuthbertson [49] and protein from Loewy [16] as modified by Lusk [15].

[4] Although their values of EeqCO$_2$ for protein, fat and carbohydrate were quoted to be essentially the same as those used by Magnus-Levy [17] it appears that the actual values used to derive their equations (table 19) were the same as those quoted by Consolazio et al. [21]. These last values are the metabolizable energy values per g food component, and not the values per g available energy which are more appropriate (see text).

[5] Brouwer [38, 50] based his values on those of Carpenter [47] which were in turn based on those given Benedict and Talbot [46]. The value for O$_2$/gNu is derived assuming Benedict and Talbot's [46] standard protein has a protein:N conversion factor of 6.25 [38, 50]. The value for fat is animal fat. For human fat, Benedict and Talbot [46] give an RQ$_{fat}$ of 0.713, and EeqO$_{2 fat}$ of 20.05 kJ/l.

[6] These authors also appear to have used the metabolizable energy values per g dietary substrates to derive the equations given in table 19 instead of gross energy values for fat and carbohydrate and metabolizable energy values for protein (i.e. instead of metabolizable energy/g available nutrient). The error is traced back to Consolazio et al. [21]. The 'adjusted' values for fat and protein (but curiously not carbohydrate) in the 'fed' state were obtained by considering the thermic effect of these nutrients. This would seem to be inappropriate since the thermic effect of food is still derived from oxidation of fat carbohydrate and protein! The values quoted by Bursztein et al. [28, 29] for the 'fed' state originate from Hunt [27].

[7] The equation given by Ben-Porat et al. [23] predicts energy expenditure using a term for urea N excretion not total urine N excretion. The O$_2$:Nu ratio given here assumes that 85% urine N is present in the form urea, as in normal man.

Table 19. Equations used to calculate energy expenditure from a fat-carbohydrate-protein oxidation mixture, and the estimated values for energy expenditure and fuel selection when 500 litres of O_2 are consumed, 425 litres CO_2 produced, and 12 gN are excreted in urine (Nu = 12 g)

Author		Equation[1]	Energy expenditure, kJ				
			total	non-protein[2]	protein[2]	carbo-hydrate[3]	fat[3]
Zuntz [25][4]		$15.97\ O_2 + 5.15\ CO_2 - 8.02\ Nu$	10,077	8,709	1,368	4,700	4,009
Loewy [16]/Lusk [15][4]		$15.97\ O_2 + 5.15\ CO_2 - 8.48\ Nu$	10,072	8,740	1,332	4,667	4,073
Magnus-Levy [17]		$15.97\ O_2 + 5.15\ CO_2 - 5.35\ Nu$	10,109	8,717	1,392	4,623	4,095
Weir [18]		$16.49\ O_2 + 4.63\ CO_2 - 9.08\ Nu$	10,104	8,717	1,332	4,473	4,299
Consolazio et al. [21][5]		$15.82\ O_2 + 4.85\ CO_2 - 12.47\ Nu$	9,821	8,772	1,284	4,518	4,019
Passmore and Eastwood [22][5]		$15.80\ O_2 + 4.86\ CO_2 - 12.0\ Nu$	9,822	8,538	1,284	4,518	4,020
Bursztein et al. [28][5]	'fasting'	$15.85\ O_2 + 4.85\ CO_2 - 12.15\ Nu$	9,840	8,556	1,284	4,531	4,026
	'fed'	$21.26\ O_2 + 0.58\ CO_2 - 0.54\ Nu$	10,870	9,537	1,284	4,838	4,699
Brouwer [38, 50]		$16.19\ O_2 + 5.00\ CO_2 - 5.94\ Nu$	10,152	8,769	1,380	4,581	4,188
Brockway [48]		$16.58\ O_2 + 4.51\ CO_2 - 5.90\ Nu$	10,136	8,756	1,380	4,498	4,256
Ben-Porat et al. [23][6]		$16.37\ O_2 + 4.57\ CO_2 - 11.88\ Nu^6$	9,984	8,760	1,224	4,622	4,138
Present paper		$15.91\ O_2 + 5.21\ CO_2 - 4.65\ Nu$	10,114	8,722	1,392	4,448	4,274

[1] O_2 and CO_2 are in litres and urinary N (Nu) is in grams.

[2] Energy expenditure from protein oxidation is calculated by multiplying gNu with the energy equivalent of 1 gNu (table 18). Nonprotein energy expenditure is calculated by subtracting the energy attributable to protein oxidation from total energy expenditure.

[3] The independent contributions of carbohydrate and fat to nonprotein energy expenditure were calculated using equation 10.

[4] These equations were not given by the original authors. They were derived (see text) using their coefficients, which are given in table 18.

[5] These equations should essentially be the same as that of Magnus-Levy [17], since the stoichiometry for carbohydrate, fat and protein oxidation is the same. The differences are to a small extent due to mathematical errors in some of their deviations, but largely to the authors inappropriately relating metabolizable energy values per g *ingested* nutrients (instead of metabolizable energy values per g available nutrients) to O_2 consumption and CO_2 production (see text and legend to table 16). The equation given by Bursztein et al. [29] in the 'fasting' state is slightly different from that given by Bursztein et al. [28].

[6] In the original equation of Ben-Porat et al. [23] urine N is given as urea N and the correction coefficient is 13.98 N. In the calculations of energy expenditure and fuel selection, the present authors have assumed that urea N accounts for 85% of total urine N as in normal man. Therefore, the correction factor for protein oxidation is 13.98 Nu $\times$ 0.85 = 11.88 Nu.

gain. However, if the energy derived from the net oxidation of fuels (or conversion of carbohydrate to fat) is known (see section on 'Estimating the Proportion of Energy Derived from the Oxidation of Different Fuels'), then the ATP gain associated with oxidation of each fuel can be calculated using the energy equivalents of ATP tabulated in table 17.

Table 20. Equations[1] for estimating total energy expenditure (kJ) associated with the oxidation of a protein fat-carbohydrate mixture from measurements of gaseous exchange alone, and calculated energy expenditure when 500 litres O_2 are consumed and 425 litres CO_2 are produced (RQ = 0.85)

Author	Equation[1]	% error[2]	Calculated energy expenditure, kJ[3]
Zuntz [25]	15.80 O_2 + 5.10 CO_2	0.7	10,068
Loewy [16]/Lusk [15]	15.79 O_2 + 5.09 CO_2	0.8	10,058
Magnus-Levy [17]	15.86 O_2 + 5.11 CO_2	0.5	10,102
Weir [18]	16.29 O_2 + 4.57 CO_2	0.8	10,087
Consolazio et al. [21]	15.55 O_2 + 4.77 CO_2	1.2	9,802
Passmore and Eastwood [22]	15.54 O_2 + 4.78 CO_2	1.1	9,802
Burszstein et al. [28] 'fasted'	15.59 O_2 + 4.77 CO_2	1.1	9,822
'fed'	21.24 O_2 + 0.58 CO_2	0.05	10,867
Brouwer [38, 50]	16.07 O_2 + 4.96 CO_2	0.5	10,143
Brockway [48]	16.45 O_2 + 4.48 CO_2	0.5	10,129
Ben-Porat et al. [23]	16.09 O_2 + 4.49 CO_2	1.2	9,953
Present paper	15.82 O_2 + 5.18 CO_2	0.4	10,112

[1] These equations, which were derived from those in table 19, assume that protein metabolism (assessed using individual author's constants for protein – see table 18) accounts for 15% of the total energy expenditure. The calculations follow the same procedure as in section IV C. O_2 and CO_2 refer to the number of litres consumed and produced respectively. The second column indicates the error that arises with each equation when the energy derived from protein oxidation accounts for 10% more or 10% less than the 15% assumed in the equation, e.g. when protein oxidation accounts for 5 or 25% of energy expenditure.

[2] % error for each 10% of energy derived from protein oxidation that is above, or below, the assumed 15% of total energy expenditure.

[3] Calculated energy expenditure (kJ) when 500 litres O_2 are consumed and 425 litres CO_2 produced. Note the similarity of tabulated values to those of total energy expenditure given in table 19, which were calculated using the same values for gaseous exchange (and 12 g urine N excretion).

The Consequences of Using Different Coefficients for Individual Fuels to Calculate Energy Expenditure and Fuel Selection

Since various authors have adopted different coefficients (RQ and $EeqO_2$) for the same fuels (table 18), different equations for calculating energy expenditure have emerged (tables 19, 20).

These equations (table 18) are derived not only from the RQ and EeqO$_2$ of protein, fat and carbohydrate, but also from the O$_2$:Nu ratio, i.e. the amount of O$_2$ used in protein oxidation per gram N excreted in urine (see above). A value for this ratio does not appear to have been given by Magnus-Levy [17] although other authors, e.g. Consolazio et al. [21] and Ben-Porat et al. [23], have calculated ratios of 6.03 or 6.04 litres O$_2$/gNu on the assumption that the protein:N conversion factor for Magnus-Levy's protein is 6.25. Magnus-Levy's 'protein' [17] was based on the composition of casein, which has a protein:N ratio (6.38) close to 6.25 [40] and Zuntz's muscle substance [25], which has a slightly lower protein:N ratio. For meat protein, Loewy [16] and Lusk [15] calculated an O$_2$ (l):Nu(g) factor of 5.94 and a protein:N conversion factor of 6.20. The value of O$_2$:Nu given by Brouwer [38, 50] and Brockway [48] is 5.98 litres/gNu and that by Livesey and Elia [7] 5.95 litres/gNu. The energy expended per gN excreted in urine (EeqO$_{2\,prot}$·O$_2$:Nu) has been estimated by various authors to be between 111 and 116 kJ.

When the same values of gaseous exchange measurements (500 litres O$_2$, 425 litres CO$_2$) and urine N (12 gN) (typical daily values in sedentary adult humans) are used to estimate energy expenditure using the various proposed equations (table 19), all estimates are within 10% of each other, or 3% if the value obtained by the 7th equation in table 19 is excluded. Furthermore, if some of the other values are discounted on the basis of incorrect assumptions used to derive the predictive equations (see footnote to table 18), then the range in results becomes much smaller ($< 1\%$). In the examples given in table 19, the calculated energy derived from fat oxidation is estimated to vary by 17% (690 kJ/day), protein oxidation 13% (168 kJ/day) and carbohydrate oxidation by 6% (252 kJ/day) (table 19). If the results obtained with the 7th equation in table 19 are discounted, the largest difference in the estimation of fat oxidation is 6% (265 kJ/day), protein oxidation 13% (168 kJ/day) and carbohydrate oxidation 3% (232 kJ/day).

The simplified equations in table 20, which do not include a term for urinary N excretion, are derived from those listed in table 19. They incorporate the coefficients for protein given by individual authors (table 18) and assume that protein oxidation accounts for 15% of energy expenditure. They produce an error (relative to the parent equation) of only 0.1–1.2% for each 10% of energy expenditure erroneously assumed to be derived from protein oxidation (table 20). If a standard equation for a fat-glucan mixture (EeqO$_2$ (kJ) = 15.913 + 5.207 RQ) is used to predict the energy expenditure associated with the oxidation of 'atypical' fuels alone the errors are generally small (-4.6 to $+6.8\%$). In practice the errors are likely to be much smaller since in biological systems the oxidation of 'atypical' fuels contribute only a proportion of total energy expenditure.

Inaccuracies may also arise when erroneous assumptions are made about the nitrogenous end products. The effect of this variation on the interpretation of gaseous exchange involving oxidation of a fat-protein-carbohydrate mixture is indicated in table 21. Here the same total gaseous exchange (500 litres O_2 consumption, 400 litres CO_2 production) is used to calculate energy expenditure and fuel selection in a mixture in which 93.0 g Kleiber's [14] standard protein (15 gN) is oxidized to different nitrogenous end products (table 20). The fate of the protein not only affects calculations about O_2 consumption and energy derived from protein oxidation, but also the quantitative interpretation of nonprotein metabolism. In the example given in table 21, the energy derived from carbohydrate oxidation is 2-fold greater when the nitrogenous end product is uric acid or allantoin than when it is ammonia. Since the nitrogenous end products vary in different species (e.g. uric acid in birds, ammonia in worms, crustacea and marine lamellobranch, urea in many mammals, and allantoin in some mammals and reptiles) it is inappropriate to universally use the same

Table 21. Analysis of a fat-carbohydrate-protein oxidation mixture which utilizes 500 litres O_2, produces 400 litres CO_2 (RQ 0.8) and is associated with the conversion of 93 g of Kleiber's standard protein ($C_{100}H_{159}N_{26}O_{32}S_{0.7}$) to various nitrogenous end products (15 gN)

Nitrogenous end product	Gaseous exchange, l				Substrate utilization, kJ			
	protein		nonprotein		protein	fat	carbo-hydrate	total
	O_2	CO_2	O_2	CO_2				
Urea	97.26	80.34	402.74	319.66	1,895	5,618	2,455	9,968
Ammonia	97.26	92.40	402.74	307.60	1,914	6,434	1,577	9,925
Creatinine	73.25	60.36	426.75	339.64	1,400	5,890	2,669	9,959
Uric acid	88.21	62.36	411.79	337.65	1,726	5,013	3,298	10,037
Allantoin	91.26	68.35	408.74	331.65	1,783	5,213	3,018	10,014

Calculations are made using the stoichiometry for the oxidation of Kleiber's standard protein (tables 6, 7), and equations 7 and 11 for assessing carbohydrate and fat utilization (nonprotein metabolism).

equations to calculate the proportion of energy derived from different fuels. Note, however, that total energy expenditure is similar in the different examples shown in table 21. This is because the $EeqO_{2\,prot}$ for the metabolism of protein to different nitrogenous end products does not vary greatly (tables 6, 7).

Assessment of the Errors

Errors Associated with the Use of CO_2 Production Alone when Calculating Energy Expenditure

There are several circumstances when measurements of CO_2 production are made without concurrent measurements of O_2 consumption. These include studies in which tracers, such as doubly labelled water, are used to estimate CO_2 production. In the labelled water method, ^2H_2O and $H_2^{18}O$ are given simultaneously to an animal or human subject and the decrease in enrichment of the isotopes is followed with time (fig. 5). The decrease in enrichment of deuterium is due to loss of enriched water from the body and replenishment with less enriched ordinary water that is drunk (i.e. the change is due to water turnover). The rate of decrease in the isotopic enrichment of $H_2^{18}O$ is more rapid than that of ^2H_2O. This is because ^{18}O is lost from the body not only as $H_2^{18}O$ but also as $C^{18}O_2$, which equilibrates with water under the catalytic activity of carbonic anhydrase.

$$H_2O + CO_2 \rightleftarrows H^+ + HCO_3^- \tag{43}$$

The difference in the rate of loss of ^2H_2O and $H_2^{18}O$ from the body represents CO_2 production (see Prentice [5] for details of calculations and correction factors). In a typical human study, CO_2 production is estimated over a 2-week period in adults and 7–10 days in young children.

An alternative method for estimating CO_2 production is the labelled bicarbonate method [2, 3], which is an isotopic dilution technique. Labelled CO_2 is infused as bicarbonate and CO_2 production calculated according to equation 44:

$$CO_2 \text{ production (mol/min)} = \frac{F \cdot \text{label infused (dpm/min)}}{\text{specific activity } CO_2 \text{ (dpm/mol)}} \tag{44}$$

The specific activity of CO_2 refers to the mean specific activity over the period of study and is estimated from measurements of acid-labile CO_2 in urine. The factor F is the fractional recovery of infused label (the incom-

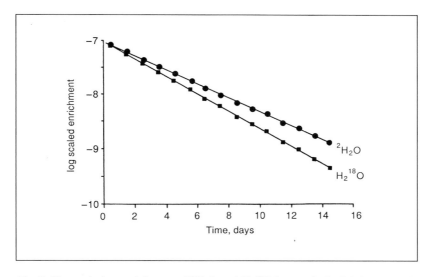

Fig. 5. Change in log enrichment of ^2H_2O and $H_2^{18}O$ in a typical adult human subject dosed with doubly labelled water (0.079 g ^2H_2O/kg body weight and 0.17 g $H_2^{18}O$/kg body weight). The difference in decline between the isotopic enrichment of ^2H_2O (●) and $H_2^{18}O$ (■) is due to production of CO_2 which equilibrates with $H_2^{18}O$ (see text).

plete recovery is due to fixation of some of the label into fat, protein, urea and carbohydrate, and entry into slowly turning over pools of CO_2).

Direct measurement of CO_2 alone (unlabelled CO_2) also has been used to estimate energy expenditure, e.g. in some ventilated patients, where the administration of high concentrations of O_2 may make it difficult to obtain accurate measurements of oxygen consumption. Therefore, the estimation of CO_2 production has made possible a prediction of energy expenditure (equation 45) in circumstances that would otherwise be too difficult to obtain reliable data. In the above three situations, energy expenditure is calculated using the energy equivalent of CO_2:

$$EE \text{ (kJ)} = CO_2 \text{ production (l)} \cdot EeqCO_2 \text{ (kJ/l)} \tag{45}$$

Unfortunately, the energy equivalent of 1 litre CO_2 ($EeqCO_2$) varies considerably, more than the $EeqO_2$ [6, 51]. Therefore, in situations where CO_2 alone is used to predict energy expenditure it is necessary to make a good choice for the $EeqCO_2$ of the body ($EeqCO_{2 \text{ body}}$). The choice is made according to the supposed contribution of various fuels to whole body energy expenditure. This can be done by considering the composition of the food eaten and change in body composition (see below).

The overall $EeqCO_2$ for a fat-carbohydrate-protein-alcohol oxidation mixture ($EeqCO_{2\,mix}$) [6, 51] is given by equation 46 or 47:

$$EeqCO_{2\,mix} = \frac{\text{Total energy expenditure}}{\text{Total } CO_2 \text{ production}}$$

$$= \frac{\text{Total energy expenditure}}{\dfrac{E_{prot}}{EeqCO_{2\,prot}} + \dfrac{E_{carb}}{EeqCO_{2\,carb}} + \dfrac{E_{fat}}{EeqCO_{2\,fat}} + \dfrac{E_{alc}}{EeqCO_{2\,alc}}} \quad (46)$$

$$EeqCO_{2\,mix} = \frac{100}{\dfrac{E\%_{prot}}{EeqCO_{2\,prot}} + \dfrac{E\%_{carb}}{EeqCO_{2\,carb}} + \dfrac{E\%_{fat}}{EeqCO_{2\,fat}} + \dfrac{E\%_{alc}}{EeqCO_{2\,alc}}} \quad (47)$$

where E_{prot}, E_{carb}, E_{fat} and E_{alc} represent the quantities of energy considered to be derived from protein, carbohydrate, fat and alcohol respectively ($E\%$ is the percentage energy), and $EeqCO_{2\,prot}$, $EeqCO_{2\,carb}$, $EeqCO_{2\,fat}$ and $EeqCO_{2alc}$ are the energy equivalents of CO_2 for these fuels (table 2). The RQ_{mix} is given by equation 48:

$$RQ_{mix} = \frac{0.835\,O_{2\,prot} + 1.000\,O_{2\,carb} + 0.710\,O_{2\,fat} + 0.667\,O_{2\,alc}}{O_{2\,prot} + O_{2\,carb} + O_{2\,fat} + O_{2\,alc}} \quad (48)$$

where $O_{2\,prot}$, $O_{2\,carb}$, $O_{2\,fat}$ and $O_{2\,alc}$ represent the oxygen consumption associated with the oxidation of protein, carbohydrate, fat and alcohol respectively, and 0.835, 1.00, 0.71 and 0.667 are the respective respiratory quotients of these fuels. Since the $EeqCO_2$ and $EeqO_2$ of individual fuels or fuel mixtures are related to the RQ ($EeqCO_2 = EeqO_2/RQ$), the RQ can be used to obtain an approximate estimate of $EeqO_2$ and $EeqCO_2$ (e.g. by modification of equation 42: EE (kJ) = 15.818 CO_2/RQ + 15.176 CO_2, where CO_2 is the measured production in litres). Parenthetically it should be noted that estimation of RQ_{mix} by equation 49, as suggested by some authors [e.g. 26], is incorrect because the $EeqO_2$ for different fuels varies:

$$100\,RQ_{mix} = (E\%_{prot} \cdot RQ_{prot}) + (E\%_{carb} \cdot RQ_{carb}) + (E\%_{fat} \cdot RQ_{fat}) + (E\%_{alc} \cdot RQ_{alc}) \quad (49)$$

where $E\%_{prot}$, $E\%_{carb}$, $E\%_{fat}$ and $E\%_{alc}$ represent the percentage of energy derived from protein, carbohydrate, fat and alcohol respectively.

The practical problem is to predict the RQ of the body, or to estimate more directly the $EeqCO_{2\,body}$ [6, 51]. One way of predicting these coefficients is from the $EeqCO_2$ of the diet ($EeqCO_{2\,diet}$) or its 'respiratory quo-

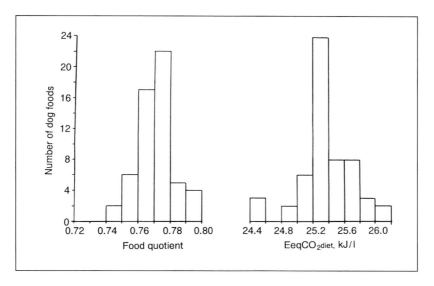

Fig. 6. Variation in the food quotient and EeqCO$_{2diet}$ for 56 different commercial dog foods. Those with the highest FQ have the lowest EeqCO$_{2diet}$. Note that dog foods (carnivorous diet) have a lower FQ and a higher EeqCO$_{2diet}$ than human diets (compare with fig. 7).

tient', which is termed 'food quotient' (FQ). Equations 46, 47 and 48 can be used to make such predictions from dietary composition data, but in this particular context equations 46 and 47 should refer to the dietary energy (not combustible energy) of the individual fuels, and in equation 48, O$_2$ should refer to the oxygen consumption necessary to oxidize the dietary fuels that are absorbed and made available to the body (the EeqCO$_2$ of the fuels remain the same – table 2). Therefore, when the body is in nutrient balance, RQ = FQ and EeqCO$_2$ body = EeqCO$_2$ diet [6] (see below for calculations when body is not in nutrient balance).

Since the composition of diets eaten by different species varies considerably, it is not surprising that the EeqCO$_{2\,diet}$ and FQ also varies considerably. The diet of carnivorous animals (e.g. cats, dogs, birds of prey), which consist mainly of fat and protein, has values of FQ and EeqCO$_{2\,diet}$ that are between the values for protein and fat, which constitute almost the whole of their diet. Figure 6 illustrates the variation in FQ and EeqCO$_{2\,diet}$ for 56 commercial dog foods. In calculating the values of EeqCO$_{2\,diet}$ and FQ it was assumed that the metabolizable energy values for dietary nutrients are

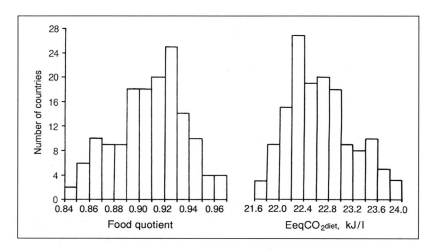

Fig. 7. Variations in the food quotient and $EeqCO_{2diet}$ for the food supplied to 147 countries in the world (dietary data from FAO of the UN [53]). The food supply which has a high FQ also has a low $EeqCO_{2diet}$.

the same as for normal man (9 kcal (37.55 kJ)/g fat, 4 kcal (16.74 kJ)/g protein or 3.75 kcal (15.69 kJ)/g available carbohydrate expressed as monosaccharide, and 7 kcal (28.29 kJ)/g alcohol [52]). The diets of birds of prey may have an even lower FQ and higher $EeqCO_{2 \text{ diet}}$ since the conversion of protein to uric acid is associated with a low RQ and high $EeqCO_2$.

Human diets have a lower $EeqCO_{2 \text{ diet}}$ than those of carnivorous animals. The $EeqCO_2$ of human diets calculated according to the principles described above generally vary from 21.8 to 24.2 kJ/l, and FQ from 0.83 to about 0.95 [6]. This variability is confirmed by dietary intake studies in various parts of the world. The $EeqCO_{2 \text{ diet}}$ of food supplied to 147 countries which cover 95% of the world population varies from 21.3 to 24.1 kJ/l, and FQ from 0.84 to 0.98 (fig. 7). There is also a large range in values for artificial feeds (fig. 8), and for the milks produced by various mammals (FQ 0.72–0.95 and $EeqCO_{2 \text{ milk}}$ 23.1–21.6 kJ/l) [6].

In some species, especially some herbivores, an end product of metabolism is hippuric acid [54], which is derived from a combination of carbohydrate and protein metabolism. This leads to further difficulties in partitioning energy expenditure to various dietary components.

Although the $EeqCO_{2 \text{ diet}}$ and FQ are important in estimating $EeqCO_{2 \text{ body}}$ and RQ of the body respectively, care should be taken when

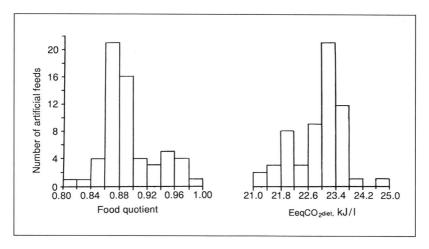

Fig. 8. Variations in the food quotient and $EeqCO_{2diet}$ for 60 artificial enteral feeds. Those with a high FQ have a low $EeqCO_{2diet}$. With permission from Am J Physiol [6].

the experimental subject or animal is not in nutrient balance, because in this circumstance nutrients are not oxidized in the same amounts or proportions as they are supplied in the diet. The extent to which $EeqCO_2{}_{body}$ and RQ_{body} are affected by dietary intake is indicated in figure 9. In constructing this figure, two important assumptions were made [see also 6]. First, it was assumed that glycogen stores do not change over the period of study. Since glycogen stores are limited in size and have little capacity for expansion, the errors that are likely to arise from this source in adult humans undertaking a 2- to 3-week doubly labelled water study are likely to be $\leq 1\%$; but they may be more in bicarbonate infusion studies carried out over 1–3 days. Second, it is assumed that protein contributes to 10% energy intake and 10% energy expenditure (see legend to fig. 9). Since protein is not the major energy source in humans, and has values for RQ and $EeqCO_2$ that are intermediate between those of fat and carbohydrate, its precise contribution to energy expenditure generally has relatively little impact on the RQ_{body} and $EeqCO_2{}_{body}$. For example, there is very little change in the results portrayed in figure 9 when it is assumed that protein contributes to 15% energy expenditure.

An implication of the first assumption (the constancy of glycogen stores) is that dietary carbohydrate given over a substantial length of time

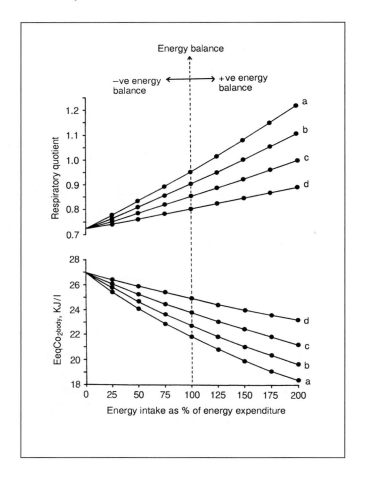

Fig. 9. The effect of energy imbalance on the respiratory quotient (RQ) and energy equivalents of CO_2. EeqCO$_{2body}$ is illustrated by a model that assumes that glycogen stores do not change and that 10% energy intake and expenditure is derived from protein. When the metabolizable energy from dietary protein is greater than 10% of energy expenditure, it is assumed that the excess energy is retained in the form of tissue protein. When the metabolizable energy of the dietary protein is less than 10% energy expenditure, it is assumed that oxidation of endogenous tissue protein makes up the deficit. Diets a–d have the following characteristics: (a) FQ = 0.95; EeqCO$_{2diet}$ = 21.87 kJ/l; (b) FQ = 0.9; EeqCO$_{2diet}$ = 22.79 kJ/l; (c) FQ = 0.85; EeqCO$_{2diet}$ = 23.83 kJ/l; (d) FQ = 0.80; EeqCO$_2$ = 24.99 kJ/l. For explanation and details see text. With permission of Am J Physiol [6].

will be utilized quantitatively, irrespective of whether the diet is given in hyper-, hypo- or normocaloric amounts. When excess carbohydrate is ingested over long periods of time it will be oxidized (or be converted to fat). Excess energy will be largely deposited in the form of fat. Under these circumstances, $EeqCO_{2\,body} < EeqCO_{2\,diet}$, and the RQ $>$ FQ (fig. 9). When the diet is provided in hypocaloric amounts, dietary carbohydrate will still be oxidized quantitatively and the energy deficit will be made up largely by oxidation of endogenous fat. Since fat has a high $EeqCO_2$ (low RQ), then the $EeqCO_{2\,body} > EeqCO_{2\,diet}$ and RQ $<$ FQ. The percentage changes in $EeqCO_{2\,body}$ and RQ_{body} with progressive degrees of over- and underfeeding are illustrated in figure 10. Diets with a low $EeqCO_{2\,diet}$ and a high FQ have the more pronounced effect on the $EeqCO_{2\,body}$ for a given degree of energy imbalance. Nevertheless, it should also be clear that the degree of energy imbalance has to be large to produce substantial errors in $EeqCO_{2\,body}$. Under- or overfeeding to the extent of 50% energy expenditure will alter the $EeqCO_{2\,body}$ and RQ_{body} by only 4–11% depending on the composition of the diet. Such degress of energy imbalance over 2 weeks (the length of a doubly labelled water study in adults) should be detectable by weighing the subject ($\pm$ anthropometry) and used to make appropriate corrections. The same principles apply to shorter studies (e.g. labelled bicarbonate studies) but assessment of energy balance from body weight/body composition is more difficult over short periods of time [6].

A physiological state of energy imbalance occurs during growth. One of the most rapid periods of growth is during the first few months of life. It is estimated that during the first 4 months of life, energy intake of a typical infant is about 34% greater than energy expenditure [6]. This increases the protein and energy 'stores' of the body and reduces the $EeqCO_{2\,body}$ by 3.2% to below that of the $EeqCO_{2\,diet}$ (i.e. $EeqCO_2$ of milk).

Errors Associated with the Use of O_2 Consumption Alone when Calculating Energy Expenditure

O_2 consumption is a better predictor of energy expenditure than CO_2 production (see fig. 3, table 2). If a constant value of $EeqO_{2\,mix}$ of 20.22 kJ/l is used to calculate energy expenditure in situations where the RQ is between 1 and 0.71, the maximum error is $< 4\%$ (e.g. using equations 18/19 or 42). A more accurate way of assessing energy expenditure is by predicting the RQ of the body or the contribution of various fuels to energy expenditure of the body (e.g. using equation 50 which is analogous to equation 47).

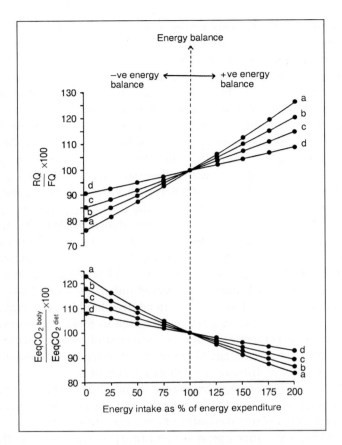

Fig. 10. The effect of energy imbalance on the ratio of respiratory quotient/food quotient × 100 (RQ:FQ × 100) (upper graph), and the ratio of $EeqCO_{2body}$:$EeqCO_{2diet}$ × 100. The four different diets are indicated a–d: (a) FQ = 0.95; $EeqCO_{2diet}$ = 21.87 kJ/l; (b) FQ = 0.9; $EeqCO_{2diet}$ = 22.79 kJ/l; (c) FQ = 0.85; $EeqCO_{2diet}$ = 23.83 kJ/l; (d) FQ = 0.80; $EeqCO_{2diet}$ = 24.99 kJ/l. The assumptions are the same as those indicated in the legend to figure 9. For explanations and details see text. With permission of Am J Physiol [6].

$$EeqO_{2\,mix} = \frac{100}{\dfrac{E\%_{prot}}{EeqO_{2\,prot}} + \dfrac{E\%_{carb}}{EeqO_{2\,carb}} + \dfrac{E\%_{fat}}{EeqO_{2\,fat}} + \dfrac{E\%_{alc}}{EeqO_{2\,alc}}} \tag{50}$$

In this equation, $E\%_{prot}$, $E\%_{carb}$, $E\%_{fat}$ and $E\%_{alc}$ are the percentage contributions to total energy expenditure of protein, carbohydrate, fat and alcohol respectively, and $EeqO_{2\,prot}$, $EeqO_{2\,carb}$, $EeqO_{2\,fat}$ and $EeqO_{2\,alc}$ are the

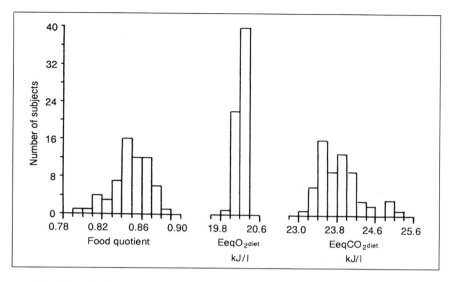

Fig. 11. Distribution of the food quotient (RQ), energy equivalent of oxygen (EeqO$_{2diet}$) and CO$_2$ (EeqCO$_{2diet}$) for the diets of 63 adults which were weighed over a period of 1 week (calculations based on individual dietary intake data) [55]. Note the percentage variation is less for EeqO$_{2diet}$ than for EeqCO$_{2diet}$ and FQ. Diets with the higher FQ tend to have higher values for EeqO$_{2diet}$ and lower values for EeqCO$_{2diet}$.

standard values of EeqO$_2$ for these fuels as shown in table 2. The principles for calculating the EeqO$_2$ of the body (and of the diet), in the presence or absence of changes in body composition, are the same as for calculating the EeqCO$_2$ of the body. The procedures are therefore not repeated here.

Since the EeqO$_2$ for different fuels is much less variable than the EeqCO$_2$, it is not surprising that when different proportions of fuels are oxidized (e.g. in a fat-carbohydrate-protein-alcohol mixture), the variation in EeqO$_{2\,mix}$ is 3- to 4-fold less than that for EeqCO$_{2\,mix}$. Therefore, had figures 5–9 been constructed using EeqO$_2$ as the variable, a 3- to 4-fold lower variability in results would have been obtained. An example is shown in figure 11, which illustrates the variability in FQ, EeqO$_{2\,diet}$ and EeqCO$_{2\,diet}$ for diets ingested by 63 individuals living in a Cambridgeshire village. The dietary intake was measured over a period of a week using the weighed food method. The percent variation in EeqO$_{2\,diet}$ between the highest and lowest value is 2.5%, whilst the variations in EeqCO$_{2\,diet}$ and FQ are 8.8 and 10.1% respectively. Since the errors associated with the use

Table 22. Percentage error that arises when equation 6 (EeqO$_2$ (kJ/l) = 15.913 + 5.207 RQ derived for a fat-carbohydrate oxidation mixture) is used to predict the EeqO$_2$ and EeqO$_2$ of various fuels

Fuel	RQ	EeqO$_2$, kJ/l		% error[2]
		true	predicted[1]	
Glycogen/starch	1.000	21.12	21.21	0
Standard fat	0.710	19.61	19.61	0
Standard protein	0.835	19.48	20.26	+4.0
Glucose	1.000	20.84	21.12	+1.3
Alcohol	0.667	20.33	19.39	−4.6
Glycerol	0.857	21.17	20.37	−3.8
Xylitol	0.909	19.92	20.65	+3.6
Sorbitol	0.920	20.89	20.70	−0.9
Maltitol	0.960	20.87	20.91	+0.2
Lactitol	0.960	20.87	20.91	+0.2
Acetoacetic acid	1.000	19.80	21.12	+6.7
β-Hydroxybutyric acid	0.889	19.89	20.54	+3.3
Acetone	0.750	19.97	19.82	−0.8
'Medium chain triglyceride'	0.739	19.69	19.76	+0.6

[1] When equation 42 is used in place of equation 6 (i.e. derived for a fat-carbohydrate-protein mixture instead of a carbohydrate-fat mixture) to predict the EeqO$_2$ of substrates, the values are 0.6% lower than those indicated in this table.

[2] The error is linearly related to the proportion of energy derived from that fuel, e.g. when only 10% of the energy is derived from a fuel and the remainder from a standard fat-carbohydrate oxidation mixture, the error is one tenth of that shown in the table.

of O$_2$ to predict energy expenditure are relatively small, even in the presence of nutrient imbalance, some workers simply report O$_2$ consumption as an index of energy expenditure.

Errors Associated with the Use of O$_2$ Consumption plus CO$_2$ Production when Calculating Energy Expenditure and Fuel Selection

The errors associated with the use of O$_2$ consumption plus CO$_2$ production for predicting energy expenditure (with or without estimates of protein oxidation – equations 7 and 19) are less than those associated with either O$_2$ consumption alone or CO$_2$ production alone. The absolute errors are small even when alcohol or atypical fuels are included in the oxidation mixture (table 22). This is largely because the EeqO$_2$ for different fuels is not very variable. For example, the $EeqO_2$ for alcohol is 20.33 kJ/l, the $EeqCO_2$ is 30.49 kJ/l, and the RQ 0.6667. At this RQ, equation 6 predicts a value for $EeqO_2$ of 19.385 kJ/l, and for $EeqCO_2$ of 29.075 kJ/l. The discrepancy between the predicted values and the actual values is only 4.64% for both $EeqO_2$ and $EeqCO_2$. Therefore, when equation 6 is used to predict energy expenditure for alcohol alone, the error is only 4.64%. When it is used to predict the energy expenditure from a mixture of fat-carbohydrate-alcohol, in which alcohol provides 10% of the energy, the error is 0.464% irrespective of the final RQ_{mix}. The errors are even smaller when some atypical fuels are included in the oxidation mixture (table 22).

Methods and examples for calculating the errors on fuel selection when there are erroneous assumptions about the nitrogenous end products (see table 21 and section on 'The Consequences of Using Different Coefficients for Individual Fuels to Calculate Energy Expenditure and Fuel Selection') and other end products of metabolism (see equation 39 and table 15), have been presented earlier. The errors in fuel selection arising from incorrect measurements of O_2 consumption or CO_2 production are discussed below.

When oxygen consumption alone or CO_2 production alone are used to predict energy expenditure, the percentage errors associated with the measurement of O_2 consumption or CO_2 production are obviously equal to the percentage error in the calculated energy expenditure. On the other hand, when a combination of O_2 consumption and CO_2 production is used to estimate energy expenditure (e.g. equation 42), the effect of an error in the measurement of one gas may be 'buffered' or even counteracted by an error in the measurement of the other gas (fig. 12, 13). However, when both O_2 consumption and CO_2 production are overestimated (or underestimated) by the same percentage, then energy expenditure will also be overestimated by the same percentage (fig. 13). Oxygen consumption has a 3-fold greater impact than CO_2 production on the estimated energy expenditure (fig. 12), as can be deduced by consideration of the constants associated with O_2 consumption and CO_2 productions in the energy equations (e.g. equation 42 and those given in tables 19 and 20). However, CO_2 has an approximately equal impact on the estimation of fuel selection as O_2 [exact impact depends on RQ (fig. 13, 14)]. A combination of errors may produce variable results (fig. 14).

Figure 15 shows the effect of incorrect measurements of O_2 consumption or CO_2 production on errors in carbohydrate and fat oxidation, expressed in relation to the actual rates of carbohydrate and fat utilization

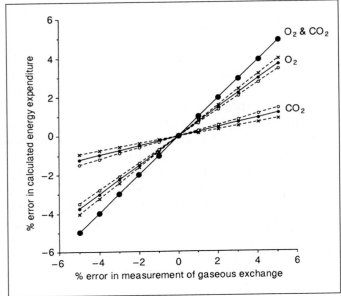

12

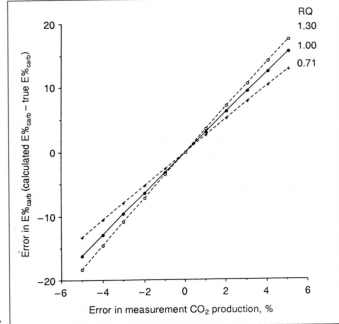

13

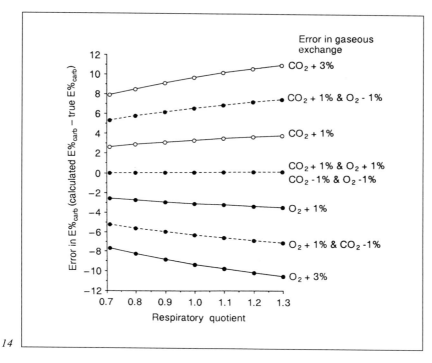

14

Fig. 12. Errors in energy expenditure resulting from errors in measurement of CO_2 production (lower), O_2 consumption (middle) and both O_2 consumption and CO_2 production (upper). The errors are calculated using equation 21 (EE (kJ) = 15.818 O_2 + 5.176 CO_2) for a fat-carbohydrate-protein mixture, which produces the same errors as equation 7 (EE (kJ) = 15.913 O_2 + 5.207 CO_2) for a fat-carbohydrate mixture. $\times$ = Results at RQ 0.71; $\circ$ = results at RQ 1.30; $\bullet$ = results at RQ 1.0; $\bullet$ = results at all RQs (when the % error in the gaseous exchange of O_2 and CO_2 is the same).

Fig. 13. Effect of incorrect measurements of CO_2 production on the errors in E%$_{carb}$ (true E%$_{carb}$ – calculated E%$_{carb}$), where E%$_{carb}$ is the carbohydrate energy utilized as % of energy expenditure – calculated using equation 7 (EE (kJ) = 15.913 O_2 + 5.207 CO_2) for a carbohydrate-fat mixture. The solid curve ($\bullet$) indicates the values at a RQ = 1.00. The dashed line with the steeper slope ($\circ$) indicates the values at a RQ = 1.30 and the dashed line with the flatter slope (+) indicates the values at RQ = 0.71.

Fig. 14. Effect of erroneous measurements of gaseous exchange on the error in % energy derived from carbohydrate utilization (E%$_{carb}$) at various respiratory quotients (fat-carbohydrate mixture). The % error in the measurement of gaseous exchange is indicated to the right of each curve. $\times$ = Error in measurement of O_2 consumption; $\circ$ = error in measurement of CO_2 production; $\bullet$ = error in measurement of O_2 consumption and CO_2 production in combination (% errors indicated). Calculated using equation 11 for a carbohydrate-fat mixture. The errors in C% are equal and opposite to those for F% (% energy derived from fat utilization – not shown).

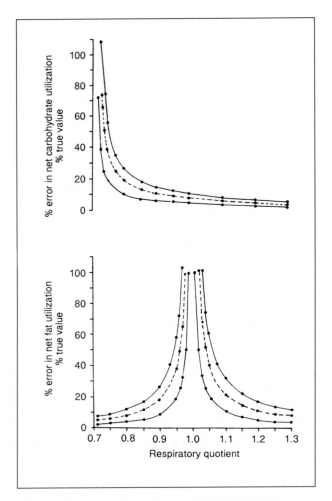

Fig. 15. The effect of overestimating CO_2 production by 1% and 3% (lower and upper solid curves on each graph respectively) on the error in carbohydrate utilization (upper graph) and fat utilization (lower graph). Results are expressed in relation to their true values:

$$\frac{100 \,(\text{difference between true and observed utilization})}{\text{true utilization}}$$

The dashed line indicates the errors associated with a simultaneous overestimation of CO_2 production by 1% and underestimation of O_2 consumption by 1%. The dashed curve indicates virtually the same error as that produced by overestimating CO_2 production by 2%. Calculations are based on equation 11 for a carbohydrate-fat mixture.

in a simple mixture of fat and carbohydrate. This contrasts with figures 12–14 which display the errors in carbohydrate and fat oxidation in relation to energy expenditure. It can be seen from figure 15 that the errors in carbohydrate utilization (as % of true utilization) approach infinity at RQ = 0.71, when carbohydrate oxidation is zero. The error becomes exponentially smaller as the RQ rises above 0.71. The errors in fat utilization (as % of true utilization) approach infinity at RQ = 1.0, when there is no fat oxidation, and become exponentially smaller as the RQ begins to deviate above and below RQ = 1.0. The curves associated with errors in O_2 consumption are essentially the same as those for errors in CO_2 production, with an opposite sign (i.e. underestimation of O_2 consumption by 1 % produces essentially a similar curve as overestimation of CO_2 production by 1 %).

The calculated percent energy derived from fat (F%) and carbohydrate utilization (C%) is affected by incorrect measurement of gaseous exchange to a greater extent than energy expenditure – approximately 4 times more with errors in O_2 consumption, and 12 times more with errors in CO_2 production. This is illustrated by comparing the magnitude of the errors in figure 12 (% error in energy expenditure) with those in figures 13 and 14 (% error in E%$_{carb}$), both of which were constructed using equation 11. Although the effects vary slightly depending on the RQ, each 1 % error in the measurement of CO_2 produces approximately 3% error in C% (and F%, since F% = 100 – C%). Percentage errors in the measurement of O_2 consumption produce virtually identical but opposite errors in C% to those arising from percentage errors in the measurement of CO_2 (see fig. 14). Therefore, the errors in C% and F% are additive when O_2 consumption and CO_2 production are erroneously estimated in opposite directions. In contrast, errors in the estimation of energy expenditure resulting from incorrect measurements of O_2 will become smaller when CO_2 production is incorrectly estimated in the opposite direction as O_2 (e.g. overestimation of O_2 and underestimation of CO_2 and vice versa) (e.g. see equations 7 and 42).

Errors Associated with the Measurement of O_2 Consumption and CO_2 Production

When the amount of CO_2 produced by a biological system is either greater or less than the amount of O_2 consumed, the volume of dry air entering the system will differ from the volume leaving the system. When the RQ is < 1.0, less CO_2 is produced than O_2 consumed and therefore the

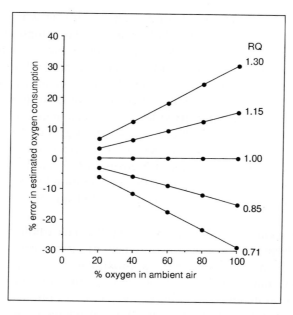

Fig. 16. The relationship between ambient O_2 concentration and the % error in O_2 consumption calculated on the erroneous assumption that the measured volume of air leaving the respiring system is equal to the volume entering the system. This assumption only applies at a respiratory quotient of 1.0, where the error is zero. The different curves indicate the error at various respiratory quotients, i.e. the O_2 concentration in the air entering the system (for details see text).

volume of outgoing air is less than the volume of ingoing air. This implies that the concentrations of gases in the outgoing air are higher than those which would exist if there was no volume change. The reverse situation occurs when the RQ is > 1.0.

When there is such a volume inequality there is scope for error in the calculations of gaseous exchange, particularly when measurements are made of only O_2 or only CO_2. However, when measurements are made of both O_2 and CO_2 concentrations, the problem of volume inequality can be taken into account by relating the measurements to the predicted concentration of N_2 ($100 - \%CO_2 - \%O_2$), which is assumed to be neither consumed nor produced by the biological system. This procedure is preferable to the assessment of volume discrepancy by direct volume measurements on ingoing and outgoing air, partly because the ingoing air is often not

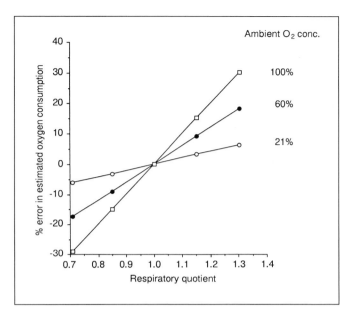

Fig. 17. The relationship between the respiratory quotient and the % error in O_2 consumption calculated on the erroneous assumption that the measured volume of air leaving the respiring system is equal to the volume entering the system. The different curves indicate the error for different ambient O_2 concentrations, i.e. O_2 concentration in air entering the system.

dried prior to entering the system (ventilation with dry air may be uncomfortable to some subjects) and partly because the discrepancy is very small and therefore difficult to measure accurately by volume air flow. For example, in an individual consuming 250 ml O_2/min and producing 212.5 ml CO_2/min (RQ = 0.8), the volume discrepancy of dry air is only 37.5 ml/min. For a subject breathing a volume of about 5 litres/min the volume discrepancy is 0.75%. If the patient's head is placed in a ventilated hood which has air flowing through it at a rate of 50 litres/min, the volume discrepancy is 10 times less (0.075%). It is obviously not practical to directly measure such small volume discrepancies.

Figure 16 shows the % error in the estimated O_2 consumption at various RQs when volume measurements in dry outgoing air are assumed to be the same as the volumes of dry ingoing air. The error is directly related to RQ. For example, for RQs ranging between 0.718 and 1.3 the errors vary from -6.1 to $+6.3\%$ when the ingoing O_2 concentration in dry air is 21%, and -29 to 30% when ingoing O_2 concentration is 100%. The errors in O_2 consumption are also directly related to the O_2 concentration in dry ambient air, which can vary considerably in artificially ventilated patients (fig. 17). At a fixed RQ the % error in O_2 consumption is independent of the rate of O_2 consumption and ventilation rate. The associated error in CO_2 production is zero when the concentration of CO_2 in ingoing air is zero and virtually zero when the concentration of CO_2 in ingoing air is close to zero.

If gaseous exchange is calculated using the volume of dry ingoing air assuming that it is the same as the volume of dry outgoing air, there is an error in both O_2 consumption and CO_2 production, but in opposite directions. The % error in O_2 consumption is less than when estimates are made on the volume of outgoing air, but the errors are now influenced by the rate of gaseous exchange.

Conclusion

Accurate determination of energy expenditure and fuel selection based on measurements of O_2 consumption and CO_2 production is dependent on the correct choice of calorimetric coefficients and equations. A rigorous examination of the principles used to derive the coefficients for fats, carbohydrates, proteins, individual amino acids and other fuels shows that a number of anomalies and errors exist in the derivation of some existing coefficients.

Although the general calorimetric equations may provide acceptable estimates for energy expenditure, with errors up to about $\pm 0-5\%$, even when unusual fuels are predominantly being oxidized, the errors in fuel selection can be magnified severalfold. With a typical fat-carbohydrate oxidation mixture, errors in the estimation of either CO_2 production or O_2 consumption leads to errors in the estimation of fat or carbohydrate utilization (as a proportion of energy expenditure) that are 3–12 times greater than the errors in energy expenditure. A magnification of such errors may also occur when the end products of metabolism differ from the usual ones.

The paper also concludes that it is possible to use indirect calorimetry to assess energy expenditure and fuel selection and deposition when there is no lipid synthesis from carbohydrate. It is also possible to modify the general equations for use in situations where there are multiple or unusual fuels utilized, or unusual end products formed.

With the use of correct measurements and 'correct' general coefficients and calculation procedures, it is estimated that the errors in energy expenditure are usually within $\pm 1.5\%$ when both O_2 consumption and CO_2 production are measured; $\pm 2-3\%$ when oxygen consumption alone is measured; and $<5\%$ to $>10\%$ when CO_2 production alone is measured, depending on the period of measurement, and on whether or not appropriate 'checks' are used. Finally, the use of a theoretical framework is recommended for optimal interpretation of measurements obtained by indirect calorimetry and tracer methods, such as the doubly labelled water and labelled bicarbonate, that primarily measure CO_2 production.

References

1   McLean JA, Tobin G: Animal and Human Calorimetry. Cambridge, Cambridge University Press, 1987.
2   Elia M, Fuller NJ, Murgatroyd P: The potential use of the labelled bicarbonate method for estimating energy expenditure in man. Proc Nutr Soc 1988;47:247–258.
3   Elia M: The estimation of short-term energy expenditure by the labelled bicarbonate method; in Whitehead RG, Prentice A (eds): New Techniques in Nutrition Research. New York, Academic Press, 1991, pp 208–232.
4   Lifson N, Gordon GB, McClintock R: Measurement of total carbon dioxide production by means of $D_2{}^{18}O$. J Appl Physiol 1955;7:704–710.
5   Prentice AM (ed): The Doubly-Labelled Water Method for Measuring Energy Expenditure. Technical Recommendations for Use in Humans. A Consensus Report by the IDECG Working Group. Vienna International Atomic Energy Agency, 1990.
6   Elia M: The energy equivalents of carbon dioxide ($EeqCO_2$) and their importance in assessing energy expenditure when using tracer techniques. Am J Physiol 1991;260: E75–E94.
7   Livesey G, Elia M: Estimation of energy expenditure, net carbohydrate utilization and net fat oxidation and synthesis by indirect calorimetry: evaluation of errors with particular reference to the detailed composition of fuels. Am J Clin Nutr 1988;47: 608–628.
8   Domalski ES: Selected values of heats of combustion and heats of formation of organic compounds containing the elements C, H, N, O, P and S. J Phys Chem Ref Data 1972;1:221–227.
9   Blaxter K: Energy Metabolism in Animals and Man. Cambridge, Cambridge University Press, 1989, p 36.
10   Livesey G: The energy equivalents of ATP and energy values of food proteins and fats. Br J Nutr 1984;51:15–28.
11   Weast RC: Handbook of Chemistry and Physics, ed 57. Boca Raton, CRC Press, 1976.
12   Ministry of Agriculture, Fisheries and Food: Intakes of Intense and Bulk Sweeteners in the UK 1987–1988. Food Surv Paper No. 29. Norwich, HMSO, 1988.
13   Livesey G, Elia M: Food energy values of artificial feeds for man. Clin Nutr 1985;4: 99–111.
14   Kleiber M: The Fire of Life: An Introduction to Animal Energetics. Huntington, Krieger, 1975.
15   Lusk G: The Elements of the Science of Nutrition, ed 4. Philadelphia, Saunders, 1928.
16   Loewy A: Oppenheimer's Handbuch der Biochemie des Menschen und der Tiere. C. Oppenheimer, Jena, 1911, vol 4, p 279.

17 Magnus-Levy A: Metabolism and Practical Medicine, capt 4: Metabolism in Man: Total Energy Exchange. London, Heinemann, 1907, pp 185–204.

18 Weir JBDeV: New methods for calculating metabolic rate with special reference to protein metabolism. J Physiol 1949;109:1–9.

19 Abramson E: Computation of results from experiments with indirect calorimetry. Acta Physiol Scand 1943;6:1–19.

20 McGilvery RW: Biochemistry – A Functional Approach, ed 2. Chapt 37: Energy Balance. Philadelphia, Saunders, 1979, pp 691–707.

21 Consolazio CF, Johnson RE, Pecora LJ: Physiological measurements of metabolic functions in man; Sect 8: The Computation of Metabolic Balances. New York, McGraw Hill, 1963, pp 313–317.

22 Passmore R, Eastwood MA: Human Nutrition and Dietetics, ed 8. Chapt 3: Energy. Edinburgh, Churchill Livingstone, 1986, pp 14–28.

23 Ben-Porat M, Sideman S, Bursztein S: Energy metabolism rate equation for fasting and post-absorptive subjects. Am J Physiol 1983;244:R764–R769.

24 Frayn K: Calculation of substrate oxidation rates in vivo from gaseous exchange. J Appl Physiol 1983;55:628–634.

25 Zuntz N: Über den Stoffverbrauch des Hundes bei Muskelarbeit. Pflügers Arch Physiol 1897;68:191–221.

26 Peters JP, van Slyke DD: Quantitative Clinical Chemistry, vol 2: Energy Metabolism, chapt 1. London, Baillière Tindall & Cox, London, 1946, pp 3–93.

27 Hunt LM: An analytic formula to instantaneously determine total metabolic rate for the human system. J Appl Physiol 1969;27:731–733.

28 Bursztein S, Saphar P, Glaser P, et al: Determination of energy metabolism from respiratory function alone. J Appl Physiol 1977;42:117–119.

29 Bursztein S, Glaser P, Trichet B, et al: Utilization of protein, carbohydrate and fat in fasting and post-absorptive subjects. Am J Clin Nutr 1980;33:998–1001.

30 Merril AL, Watt BK: Energy Values of Foods: Basis and Derivation. USDA Handbook 74. Washington, US Government Printing 1973.

31 Lentner C (ed): Geigy Scientific Tables. 1. Units of Measurement, Body Fluids, Composition of the Body, Nutrition. Basel, Ciba-Geigy, 1981.

32 Hutchens JO: Heat of combustion enthalpy and free energy of formation of amino acids and related compounds; in Handbook of Biochemistry. Boca Raton, CRC Press, 1975, pp 87–89.

33 Tzuzuki T, Harper DO, Hunt H: Heats of combustion. 7. The heat of combustion of some amino acids. J Phys Chem Ref Data 1958;62:1594.

34 Shulz AR: Computer-based methods for calculation of available energy of proteins. J Nutr 1975;105:200–207.

35 Rawitscher M, Wadso I, Sturtevant JH: Heats of hydrolysis of peptide bonds. J Chem Soc 1961;83:3180–3184.

36 Sober HA: CRC Handbook of Biochemistry and Selected Data for Molecular Biology. Oxford, The Chemical Rubber Co, 1968.

37 Wilhout RC: Selected values of thermodynamic properties; in Brown HD (ed): Biochemical Microcalorimetry. London, Academic Press, 1969, pp 305–317.

38 Brouwer E: On simple formulae for calculating the heat expenditure and the quantities of carbohydrate and fat metabolised in ruminants from data on gaseous exchange and urine-N. Eur Assoc Anim Prod 1958;8:182–191.

39 Erwin Voit and Krummancher, quoted by Magnus Levy A: Metabolism and Practical Medicine. London, Heinemann, 1907, p 193.

40 Brody S: Bioenergetics of growth, chapt 12: Methods in Animal Calorimetry. Baltimore, Reinhold Publ Corp/Waverley Press, 1945, pp 307–351.

41 Elia M, Livesey G: Theory and validity of indirect calorimetry during net lipid synthesis. Am J Clin Nutr 1988;47:591–607.

42 Atwater WO, Benedict PG: An experimental inquiry regarding the nutritive value of alcohol. National Academy of Sciences. 6th Mem. 1902;8:231–397.

43 Atkinson DE: Adenine nucleotides as universal stoichiometric metabolic coupling agents. Adv Enzyme Regul 1971;9:207–219.

44 Livesey G: Mitochondrial uncoupling and isodynamic equivalents of protein, fat and carbohydrate at the level of biochemical energy provision. Br J Nutr 1985;53: 381–389.

45 Elia M: Organ and tissue contribution to metabolic rate ; in Kinney JM (ed): Energy Metabolism: Tissue Determinants. New York, Raven Press, 1992, pp 61–80.

46 Benedict FG, Talbot FB: The Gaseous Metabolism of Infants with Special Reference to Its Relation to Pulse Rate and Physical Activity. Washington, Carnegie Institution of Washington, 1914, Publ No 201.

47 Carpenter TM: Tables, Factors and Formulas, for Computing Respiratory Exchange and Biological Transformations of Energy. Washington, Carnegie Institution of Washington, 1948, Publ No 303C.

48 Brockway JM: Derivation of formulae used to calculate energy expenditure in man. Human Nutr Clin Nutr 1987;41C:463–471.

49 Cathcart EP, Cuthbertson DP: The composition and distribution of fatty substances of the human body. J Physiol 1931;72:239–360.

50 Brouwer E: On simple formulae for calculating the heat expenditure and the quantities of carbohydrate and fat oxidised in metabolism of men and animals from gaseous exchange (oxygen intake and carbonic acid output) and urine N. Acta Physiol Pharmacol Neerl 1957;6:795–802.

51 Elia M: Converting carbon dioxide production to energy expenditure; in Prentice AM (ed): The Doubly-Labelled Water Method for Measuring Energy Expenditure. Technical Recommendations for Use in Humans. A Consensus Report by the IDECG Working Group. Vienna, International Atomic Energy Agency, 1990, pp 193–221.

52 Paul AA, Southgate DAT: McCance and Widdowson's The Composition of Foods, ed 4. London, HMSO, 1978.

53 FAO of the UN: 1986 FAO Production Yearbook, vol 40: FAO Statistics Series No 76, 1986, pp 245–251.

54 Blaxter K, Wainman FW: The utilization of energy of different diets by sheep and cattle for maintenance and for fattening. J Agric Sci 1964;63:113–118.

55 Bingham S, McNeil SNI, Cummings JH: The diet of individuals: A study of a randomly chosen cross section of British adults in a Cambridgeshire village. Br J Nutr 1981;45:23–35.

M. Elia, MD, Dunn Clinical Nutrition Centre, 100 Tennis Court Road, Cambridge, CB2 1QL (UK)

Simopoulos AP (ed): Metabolic Control of Eating, Energy Expenditure and the Bioenergetics of Obesity. World Rev Nutr Diet. Basel, Karger, 1992, vol 70, pp 132–170

The Bioenergetics of Obesity Syndrome

Julius O. Olowookere

Biomembranes and Bioenergetics Research Laboratory,
Department of Biochemistry, Ogun State University, Ago-Iwoye, Nigeria

Contents

Introduction

Precisely how obese subjects regulate cellular energy reserves, or fail to, in the face of ordinary stresses and ambient temperature changes has been identified as a major problem in the pathogenesis and bioenergetics of obesity syndrome. The basic cause of dietary obesity is an intake of calories in excess of energy expenditure. The reasons for the defective energy metabolism in obesity vary widely as there are many underlying causes. Apart from dietary obesity, other forms of obesity exist. Obesity is better studied now as a syndrome with different aetiologies. Mayer [1] has however shown that the various forms of obesity are directly or indirectly linked with deregulation of food intake. With the current biochemical and physiological knowledge in the aetiology of obesity, it is appropriate to distinguish among the genetic, hypothalamic, endocrine and psychological factors in obesity syndrome.

Perhaps more crucial to the issue of aetiology of obesity syndrome is the issue of oxidative energy production. It has now been shown that the process of taking in of oxygen is made more difficult in obese subjects, hence there are respiratory difficulties in obesity. Excessive adipose tissue coupled with excess load of fatty tissues carried on the chest wall have been implicated in the respiratory impairment. Obese subjects have the concomitant problem of having the whole body defectively oxygenated [2].

In view of respiratory difficulties in obese subjects, they usually have diminished exercise tolerance and higher resting basal metabolism. There is a growing body of evidence that a thermogenic defect plays a role in the obesity syndromes including the genetically transmitted obesity (ob/ob) [3].

Table 1. Relationship between body weight and life expectancy of a 25-year-old male [data from 8, 9]

Percent excess weight	Expected age at death
0	76
30–60	63
100	52

Jequier and Shutz [4] have established that obese animals exhibit both hyperphagia and metabolic efficiency; yet they have low work tolerance.

Beaton [5] has shown that obesity is a condition associated with a well-established risk of morbidity and mortality. In a well-illustrated paper titled 'Multidisciplinary approach to adult obesity therapy', Blackburn and Greenberg [6] highlighted the risk factors in obesity. These authors related obesity to other pathological conditions such as diabetes, hypertension, hyperlipidaemia, gout, lower back strain, arteriosclerosis, hernia and low work tolerance. It has been documented [4] that although the prevalence of obesity is quite high in most industrialized countries, and as such, it is generally recognized as a public health problem, data from developing countries suggest that the prevalence of obesity is gradually increasing and may be surprisingly high among segments of the population. The extent of morbidity and mortality in obesity syndrome is shown in table 1.

It has been observed that there are inherent limitations in the use of the various existing methods clinically adopted for the assessment of the energy metabolism of obese subject [4, 10, 11]. The need for better techniques, devoid of ambiguous interpretations, during rehabilitation and management of obesity has been advocated [12]. The metabolic dependence on mitochondrial function has been well established [13]. The biochemical properties of the mitochondrion as the power plant of the living cell coupled with its genetic peculiarity being semi-autonomous, even as a subcellular organelle, has made it a veritable marker in the elucidation of the variability in energy metabolism and thermogenesis during malnutrition. This assertion has been put to test, and findings of experiments from various laboratories have proved very informative in a parallel seasoned review titled 'Bioenergetics of protein-energy malnutrition syndrome' [14].

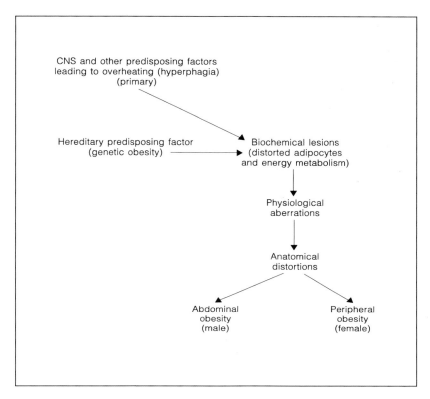

Fig. 1. A generalized scheme showing the pathogenesis of obesity syndrome in man [15–17].

In view of the variations in the aetiology of obesity, the thermogenic defects implicated in its pathogenesis, the difficulties in respiration during obesity, the low energy expenditure in relation to size (fig. 1), the observation of hyperglycaemia and low glucose uptake, the higher basal metabolic rate (BMR) in obesity and the diet-induced thermogenesis (DIT); it is pertinent to describe the current state of the art on the bioenergetics of obesity syndromes in particular. This review mainly highlights the fundamentals of bioenergetics as it relates to obesity syndromes simulated using animal models. This review also incorporates human bioenergetics data succinctly juxtaposed with the clinical diagnosis and prognosis correlates of obesity syndromes.

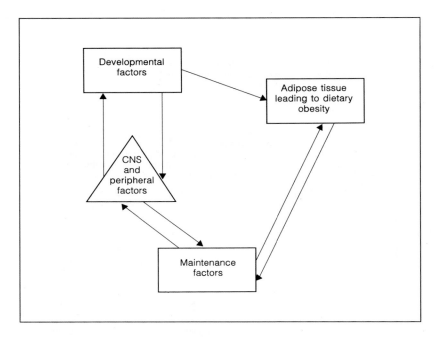

Fig. 2. The interactions of developmental and maintenance factors with nervous system in the aetiology of obesity [18].

The Problems of Obesity

Aetiology of Obesity

The various forms of obesity, in both man and animals, share the common characteristic of increased adipose tissue mass [18]. The increase in mass may be of a mild to extreme extent and may be developmental in nature or selectively affect specific adipose tissue depots. There is, however, considerable agreement in two general theories related to the aetiology of obesity:

(1) *Primary failure in the regulation of ingestive behaviour:* This theory postulates that obesity may arise primarily due to failure to regulate ingestive behaviour at the cognitive (brain) level (fig. 1). The loss of this cognitive regulatory mechanism naturally leads to hyperphagia resulting in adiposity as a result of the excessive caloric intake. It has been documented that the vast majority of notoriously unsuccessful weight control measures

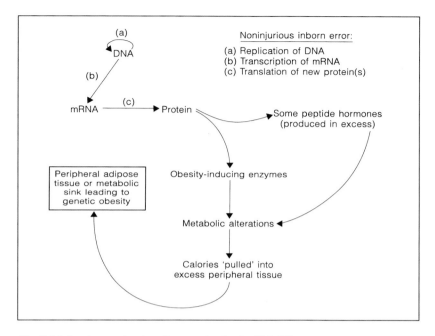

Fig. 3. Molecular mechanism for genetic obesity [15–17].

are predicated on this theory [18], since the fundamental issue of aberration in cognitive behaviour of these obese subjects is not directly treated, whilst the 'effect' which is obesity is treated; thus, the 'cause' which is the cognitive aberration is left untouched (fig. 2).

(2) *Genetic or inborn metabolic errors:* This theory postulates that some forms of obesity may arise as a result of genetic aberration leading to inborn errors and a metabolic condition whereby calories are actually 'pulled' into peripheral tissues which act as a metabolic sink. This genetic obesity, which originates at the level of deoxyribonucleic acids (DNA), is transmissible to offsprings. It is also the primary cause of metabolic alterations, plasticity and reversibility culminating in the abnormal growth and development of adipose tissue and those factors which maintain and encourage the stable obese condition. The introduction, in 1961, of the genetically obese Zucker rat (fafa) has facilitated considerable research on the developmental antecedents of the genetic obese conditions.

Figure 3 shows the molecular mechanism for the aetiology of genetic obesity. Using biopsy techniques followed by retrospective identification of the phenotype (in genetic obese subjects), Lavau et al. [19] have studied the developmental changes in plasma metabolites. It has been established [18] that the adipose tissue enzymes, including lipoprotein lipase (LPL), are elevated in genetically obese rats in consonance with the postulated mechanism (fig. 3) [15–17]. Lavau et al. [19] also observed significant elevation in adipose tissue enzymes in two depots (inguinal and dorsal scapular) as well as elevated lipogenesis from de novo lipogenic enzymes.

It is clearly evident from the account of Gruental et al. [20] for an early 'pull' of nutrients into adipose tissue during the suckling period when the baby rats are not hyperphagic. It is now clear that genetic obesity develops in the absence of hyperphagia and consequently is not a result of increased ingestion. Hitherto, Turkenkopf et al. [21] asserted that the exact mechanisms for genetic obesity in Zucker rats are still unknown. They however agreed that this model of obesity seemingly originated at the earliest postnatal stages and may even be indicated prior to birth [21]. In view of the current lack of any mechanistic explanation vis-à-vis copious biochemical data on genetic obesity, this author has hereby proposed in figure 3 a molecular mechanism for the biogenesis of genetic obesity [15–17].

Classification and Characterization of Obesity

The need to classify and characterize human obesities emanates primarily from the morphologic and metabolic variables, behavioural indicators, energy balance and the various risk factors associated with the syndrome [22]. There have been numerous attempts to classify human obesities [23–27]. As early as 1900, von Noorden proposed a dichotomous classification of *exogenous obesity* and *endogenous obesity*. This, essentially, is the same as *dietary obesity* and *genetic obesity* [28]. This classification lumped the majority of obese subjects into the exogenous group while a few individuals come under the endogenous group. The practical utility of the above classification was very limited. Mayer [1] also proposed another dichotomous classification, namely *regulatory obesity* and *metabolic obesity*. More recently, Sjorstrom [29] proposed a classification based upon adipose cellularity, namely *hypercellular obesity* and *metabolic obesity*.

As a result of all these classifications, the literature is often confusing and seemingly contradictory. In regard to obesity syndrome, Harrison [30] has adduced four main reasons for developing classifications. These are: (a) to improve understanding; (b) to improve clinician prognostic ability; (c) to improve researchers' and clinicians' ability in matching appropriate preventive and treatment strategies to individuals and groups, and (d) to describe the population of obese patients or subjects in functional terms of significance to the individual's physical and mental well-being.

Each purpose will require somewhat different data in spite of the obvious overlap in the database that would be necessary for all four types of classifications and may be more useful in distinguishing the various forms of obesity. Listed below (table 2) are four distinguished groups based on rare pathophysiological syndromes associated with obesity.

Features of Dietary Obesity

Hyperphagia has been identified as a causative factor in dietary obesity. The distinctive ingestive behaviours, such as unrestrained eating, have been described by Stunkard [31]. The following other features have also been documented for diet-induced obese subjects: (a) increases in BMR (fig. 4);

Table 2. Characterization of obesity based on pathophysiological syndromes

Endocrine abnormal disorders	Central nervous disorders	Chromosomal syndromes or congenital with:	
		anomalies	fat distribution
Excess adrenocorticosteroids, Cushing's disease, corticosteroid medication	Trauma or surgical injury	Prader-Labhart-Willi syndrome	Steatopygia
	Tumour	Laurence-Moon-Bardet-Biedl syndrome	Partial lipoatrophy with secondary lipohypertrophy
Excess insulin (e.g. 15–20%) of patients with insulin-producing tumors	Infiltrative lesions, e.g. leukaemia histiocytosis $\times$ sarcoidosis	Carpenter's syndrome	Madelung's neck/ Launo-Bensaude syndromes
Hypogonadal syndrome (e.g. Klinfelter's syndrome, Kallmann's syndrome)	Post-viral encephalopathies (Kline-Levin syndrome)	Alström's syndrome	
		Pseudohypoparathyroidism	
		Down's syndrome	
		Börjeson-Forssmann-Lehmann syndrome	
		Beckwith-Wiedemann syndrome	

(b) increases in DIT (fig. 4); (c) increases in resting energy expenditure; (d) excessive or abnormal growth due to adipose tissue accumulation; (e) metabolic derangement, and (f) defective energy transducing mechanism.

However, when DIT was expressed as a percent of the energy intake (each subject received 41.5 kcal/kg fat-free mass over 24 h), it was found that DIT was significantly ($p < 0.001$) lower in the obese women than in the lean controls as revealed in figure 5.

These data are on the concept of a reduced thermogenic response in obese women with a history of childhood onset of obesity (fig. 5) whereas the overall picture as revealed in figure 4 showed that the energy expended in the postprandial state is higher in the obese than in the lean controls. The present results revealed that in spite of a reduced DIT, obese women have a greater postprandial energy expenditure than the lean controls. Here, therefore, lies a puzzle for researchers in cellular bioenergetics which may lead to ambiguous interpretation of the energy status in obesity syndrome. No wonder that Jequier and Shutz [4] asked the controversial (yet pertinent) question in their classical paper titled 'Does a defect in energy metabolism contribute to human obesity?'

Regional Differences in Adipocyte Metabolism in Obesity

It has been documented [32] that there are regional differences in the distribution of adipose tissue in obese men and women. The same author implicated the sex hormones in adipocyte metabolism in dietary obesity. It has been demonstrated that abdominal obesity is more closely correlated with the metabolic aberrations of the obese state than peripheral obesity. The fact that women have preferential peripheral depots while men have preferential abdominal depots supports this hypothesis.

It has been demonstrated that the abdominal depots are more responsive to lipolytic hormones than the peripheral depots [32]. These findings suggest that an expansion of the abdominal depots would lead to enhanced fatty acids release, compared to peripheral depots. Since abdominal obesity is closely related to metabolic aberrations [32], the increase in fatty acids may be due to the enhanced lipolysis of abdominal cells. It has been documented that fatty acids are the link between obesity and carbohydrate intolerance. It is well known that obese individuals are hyperinsulinaemic, insulin-resistant and have an increased propensity for diabetes (table 3). It has been shown in animal studies [33a] that during pregnancy and lactation, certain adipose depots may be preferentially degraded, probably to supply the mammary glands with fatty acids.

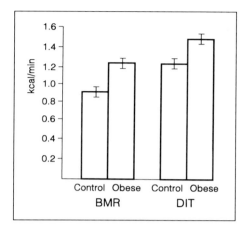

Fig. 4. BMR and DIT measured over 24 h in 20 obese and 8 lean control women. BMR: p < 0.01; DIT: p < 0.001 [data from 4].

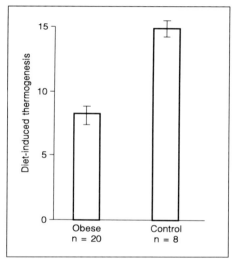

Fig. 5. DIT measured over 24 h in 20 obese and 8 lean control women. DIT is calculated as the ratio between the integrated increase in resting energy expenditure over BMR and the energy content of the meals. p < 0.001 [data from 4].

The elevated fatty acids provide the link between obesity and hyperinsulinaemia by reducing insulin clearance [32] which also provides a clue for the association between obesity and reduced glucose tolerance. That regional distribution of adipose tissue varies between men and women has been shown to be influenced by sex hormones [33b]. The regional differences in adipocyte metabolism may contribute to the intricate energy problems and metabolic aberrations in obesity syndrome.

Table 3. The major clinical correlates of a 'central', 'abdominal' and/or 'upper body' type of obesity which have been observed in various groups of subjects [from 30]

Type of obesity	Clinical correlates
Central or abdominal	Diabetes, glucose intolerance, hyperinsulinaemia
Central or abdominal	Hypertriglyceridaemia
Central or abdominal	Gout
Central or abdominal	Coronary heart disease
Central or abdominal	Urinary calculi
Central or abdominal	Advanced maturation

Treatment of Obesity

There are many approaches to the treatment of obesity. The methods listed below represent a condensation of the various techniques currently being used to treat obesity in man: (a) dietotherapy [34]; (b) pharmacologic approaches to the regulation of metabolism and obesity [35]; (c) central mechanism of anorectic drugs in the treatment of obesity [36]; (d) thermogenic agents as a treatment of obesity [37]; (e) surgical approach to the treatment of obesity [38]; (f) psychological approach in the treatment of obesity [39]; (9) integration of current modes of management; (h) weight control programmes, and (i) physiological prognostic approach in the treatment of obesity [40].

The spectrum of treatment listed above carries along its trail both in terms of their respective advantages and disadvantages. The various forms of obesities make the appropriate choice of treatment a primary assignment in the management of obesity.

Generally, the biochemistry of lipid biosynthesis and utilization are implicated at the sites at which regulation by pharmacologic modulation might aid in the treatment of obesity. The key steps involved in lipid and carbohydrate metabolism, which may be amenable to pharmacologic modification, hence cellular bioenergetics of obese subjects, include: (a) absorption of dietary fat and carbohydrate; (b) hepatic lipogenesis, lipoprotein synthesis and secretion; (c) lipoprotein lipase-induced lipolysis of circulating triglycerides and subsequent triglyceride synthesis in adipocytes; (d) fatty acid mobilization by hormone-sensitive lipase, and (e) lipid and carbohydrate oxidation.

Whatever the method of treatment adopted, the energetics of the cell could not be affected. Figure 6 shows the sites of pharmacological modu-

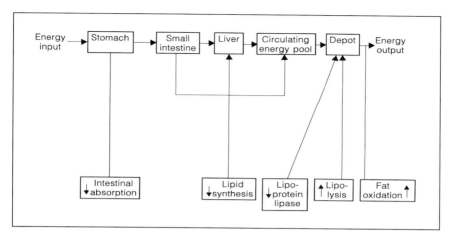

Fig. 6. Sites of pharmacological modulation of lipid and carbohydrate metabolism for obesity therapy (vis-à-vis cellular bioenergetics).

lation of lipid and carbohydrate metabolism for obesity therapy. The relationship between these 'sites of action' as well as the 'effects of other methods of treatment of obesity' vis-à-vis circulating energy pool becomes yet another factor which has necessitated the focus in this review on the bioenergetics of obesity syndrome.

Hormones of Significance in Obesity Syndrome

Several pieces of evidence point to a CNS source for control of thermogenesis in genetically obese rodents [41]. The fact that some of the problems can be ameliorated by the removal of glucocorticoids by adrenalectomy implicated hormonal changes in genetic obesity. Figure 3 on the molecular mechanism for genetic obesity supports this observation.

Figure 7 shows defects in genetically obese rodents which if corrected by adrenalectomy (fig. 7b) are corticosteroid-dependent and (fig. 7a) corticosteroid-independent if not corrected by adrenalectomy [23]. It has been shown that abdominal depots are more responsive to lipolytic hormones. The finding that abdominal depots are found in males while peripheral depots are found in females suggests that there are sex differences in adipose tissue distribution in obesity. These differences in tissue distribution are seen early in life and it is maintained even in grossly obese subjects that the sex hormones play an important role even in nongenetic obesity.

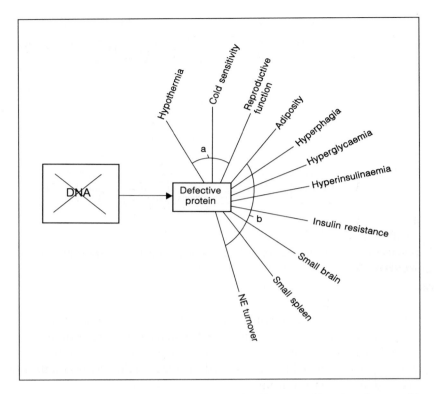

Fig. 7. Defects in genetically obese rodents which if corrected by adrenalectomy (b) are Corticosteroid-dependent and (a) corticosteroid-independent if not corrected by adrenalectomy [23].

Anthropometric Differences and Their Clinical Correlates in Obesity

Description of actual measured thickness of subcutaneous fat, usually from anthropometry or measurement of soft tissue radiographs, has been characterized by several different approaches. The term 'absolute' and 'relative' fat patterning as well as the effects of diets on adipose tissue as suggested by Steingrimsdottir et al. [42] forms the basis of the anthropometry of obesity. Absolute fat patterning refers to the distribution of subcutaneous fat in actual thickness as measured with calipers, radiography or other techniques.

The loading of individual skinfolds on a component can be used to interpret the biological meaning of the component, since skinfold thickness measurements from different sites in the same individual show significant positive intercorrelation. The pattern of relative fat distribution appears to be minimally variable in the same individual, at least in adulthood, even in the face of overall fatness. Thus, anthropometry has the potential for prediction of obesity type even in the pre-obese type. The most consistent of these observations are those related to noninsulin-dependent diabetes glucose intolerance, hyperinsulinaemia, hyperlipidaemia and coronary heart disease. The hypothesized relationship of obesity with gout and urinary calculi are tenuous, whilst only one study has so far supported the fact that accelerated maturation is a clinical correlate of obesity [43]. Studies by Stallones et al. [44] have shown that there is no relationship between blood pressure (hypertension) and fat distribution independent of body weight.

The relationship, if any, between variability in developmental patterns of subcutaneous fat distribution and possible development of obesity and related clinical pathologies have not yet been explored, thus there is need for further research in this crucial clinical area to debunk some generalizations and/or speculations.

Psychological, Behavioural and Environmental Factors in Obesity Syndrome

The psychological (or better referred to as cognitive) factors have been categorized into three domains in obesity syndrome:

(1) *The 'restraint ingestion concept':* This concept studied most extensively by Herman and Polivy [45] refers to a person's resistance to pressures to eat, either psychologically or physiologically. This may be due to the individual defending a biological 'set point' [46]. One pertinent issue here is the fact that restrained eaters share several cognitive characteristics, including perfectionistic expectations about dieting. Invariably and surprisingly, restrained eaters show negative emotions to a 'slip' during treatment of obesity. It has been shown [47] that obese individuals with 'restrained cognitive profile' show poor prognostic profile for treatment.

(2) *The abstinence violation factor:* This effect was vividly described by Cummings et al. [48]. Essentially if this factor comes to focus when a dieter has established self-imposed rules for diet programme (e.g., 'I will never eat chocolate cake again'). Deviation from such self-imposed rule can result into full-fledged relapse, if the person feels that he or she has violated the abstinence rule. Most professionals, who work with obese patients, still

feel that this technique has some merit, if backed up with counselling and elimination-substitution technique.

(3) *The psychological cognitive factor:* This refers to an individual's commitment to institute self-regulatory action if weight exceeds a self-determined criterion level. With this concept, which tries to 'nip in the bud', the weight increase threshold has been found to produce positive results [7].

Behavioural and Environmental Factors in Obesity

One behavioural factor which predicts the likelihood that a person will be obese is physical activity. Epidemiological data show a strong relationship between inactivity and obesity [49, 50].

The prevalence of obesity in the USA has doubled since 1900 despite decreases in average daily energy consumption, presumably because of the substantial decreases in energy expenditure [47]. The second behavioural factor is a person's ability or willingness and determination to engage in self-enforcement. It includes the ability of the individual to recognize and acknowledge positive weight restrictive measures. This, as it were, is an internal source of control, and subjects with this ability seem better able to adhere to a long and arduous diet programme.

Environmental component of obesity syndrome largely revolves around the social support given to an individual in the environment which is related to susceptibility to aid recovery from a variety of serious diseases [51]. Family history is strongly associated with an individual's chance of being obese [52]. This could be an expression of a physiological and cultural environment. It may even be as a result of the environment on the 'genes' of obese parents. Whatever the mechanism of the environmental factor may be, it has been established that family history is a useful factor in the diagnosis and prognosis of obese subjects.

Effects of Exercise on Food Intake

The effect of exercise on food intake in overweight people has not produced unequivocal results. It has been shown that prescribing exercise programmes without energy restriction in obese men and women [53] invariably leads to change in their eating habits. This may occur consciously or unconsciously. It has been experimentally demonstrated [54] that obese subjects, whether they exercised or not and how much they exercised, seemed to make very little difference in their food intake behaviours. However, enhancing the quality of the food significantly increased the energy intake and made energy balance positive. The crucial issue of concern to bioenergeticists is the resultant difference between energy expenditure and energy intake with or without exercise in persons with obesity syndrome.

Mechanism of Cellular Energy Transduction

Mitochondrion as Power Plant of the Cell

Mitochondria are subcelluar double membranous organelles. Their main function is the generation of adenosine triphosphate (ATP) at the expense of the electron transport to oxygen. The mitochondrion has been described as the 'power plant' of all eukaryotic cells. The energy-rich ATP molecules (produced by these organelles) are used to provide the cellular energy required for all physiological and biochemical functions [13]. Mitochondria also participate in many other metabolic functions such as the tricarboxylic acid cycle [55] and the oxidation of fatty acids resulting from the biodegradation of acylglycerols (triglycerides) and phospholipids. Most of the enzymes involved in these crucial metabolic reactions are compartmentalized inside the mitochondrion. The flux of metabolites (derived essentially from dietary intakes) across the mitochondrial membrane in both directions is regulated by specific transport systems localized in the membranes. The inner membrane is different functionally from the outer membrane [56]. There are many points of contact between the outer and the inner mitochondrion membrane [57]. The mitochondrion also contains a semi-autonomous genetic apparatus which is distinct from the nuclear genome. This, perhaps, is one of the most pronounced functional uniqueness of mitochondria. It has been documented that the mitochondria synthesize about 15% of their own proteins translated on the mitochondria genome [58]. It has also been shown [59] that the transcriptional and translational abilities of mitochondria are age-dependent. Olowookere and co-workers [14, 28, 60–63] have shown that mitochondrial energy-transducing capacities are affected by low dietary protein intake.

Mitochondrial Coupling Membranes

Perhaps it is most pertinent to state the fact that the mitochondria have double membranes. This feature, apart from the semi-autonomous genome, is another peculiarity of this unique organelle. There is ample evidence to support the fact that the morphology of plant mitochondria is very similar to that of animal mitochondria [64–68]. The respiratory chain of higher plant mitochondria has been shown to resemble that in animal mitochondria in many respects [69]. It has similarly been documented that the chloroplasts are morphologically related to the mitochondria by having double membranes.

The fact that chloroplasts and the mitochondria are primiarily and secondarily involved in ATP generation (during photophosphorylation and oxidative phosphorylation) point to a common evolutionary line. Functionally, the topology and the organization of component smaller molecules, e.g. vitamin A and membrane bound enzymes in the inner mitochondrial membrane, have been explained in a new molecular model for biomembranes [70].

Thermodynamic Aspects of Cellular Energy Transduction

Of more relevance to the subject of mitochondrial energy transduction is the second law of thermodynamics. This law, in its simplest form, states that 'No spontaneous transformation of energy to another is ever 100% efficient'. This is because there is bound to be some loss of energy (as heat), and means that the energy-transducing capacity of mitochondria in an *isothermal* medium, such as the cell, is not 100% efficient. It has been shown that the efficiency of mitochondrial energy transduction is only 48% in man.

The heat dissipated is used to maintain the cellular homeostasis of the internal milieu within a narrow range of temperature change, which is 37 °C in man. The roles of the numerous enzymes in the inner mitochondrial membrane ensure the success of this biological energy transformation. The mathematical relationship of the thermodynamic components to the other vital energy indices is discussed in the succeeding section.

Chemiosmotic Theory of Energy Transduction

The term 'chemiosmotic' is coined from two words, namely *chemical* and *osmotic*. The most acceptable theory of energy transduction has been shown to have both chemical and osmotic components [71].

According to Mitchell [71], who is the proposer of the chemiosmotic theory, the mechanism of energy transduction in the mitochondria could be explained simply, thus that the components of the respiratory chain are vectorially organized in the mitochondrial inner membrane; that the electron transport is accompanied by the translocation of protons outwards across the inner membrane; that due to the low ion conductance of the membrane, the free energy change involved in the electron transfer reaction generates a radient of electrochemical gradient across the inner membrane. The potential difference is a measure of the energy pool which can be utilized by the energy-requiring mitochondrial processes.

It is useful here to explain further that 'the function called the "membrane potential" in electrophysiology is really the electric membrane potential, which would generally be part of the protonic potential' [72]. When quantitatively related, the various components of 'membrane potential' shed light on the energy transduction processes thus:

$$\Delta p = \Delta \psi - Z \Delta pH,$$

where $\Delta \psi$ = electrical membrane potential; Δp = protonic membrane potential, and $- Z \Delta pH$ = chemical or thermodynamic parts of membrane potential.

From the above, it would be appreciated that 'membrane potential' could be *electric* or *protonic*. The protonic component of the 'membrane potentials' is however greater than the electric membrane potential by the value equivalent to $- Z pH$. This difference represents the chemical or thermodynamic component of the membrane potential as shown in the equation above. However, it has been established [72] that the H^+ concentration/gradient, which gives rise to the protonic membrane potential, represents the osmotic component, and, indeed, the main driving force in the formation of ATP. It has therefore been firmly established [72] by the chemiosmotic theory that the oxidative and the photosynthetic systems are bioenergetically 'coupled' by protonic-derived force.

Effects of Different Levels of Dietary Intakes on the Metabolic Functions of Mitochondria

The incidence of severe hypoglycaemia, which may be accompanied by hypothermia during protein-energy malnutrition (PEM), has been identified as the most obvious sign of an impaired carbohydrate metabolism in the PEM-diseased state [73]. It has been shown that kwashiorkor patients cannot utilize intravenously administered glucose [74]. Kwashiorkor patients have slightly reduced fasting blood sugar and diminished glucose tolerance [75]. There is an increase in serum lactate and lactate dehydrogenate (LDH) during the acute phase of kwashiorkor [76]. Elevation of lactate, the end-product of anaerobic metabolism, suggests that glucose metabolism passes freely through the glycolysis; pyruvate is not being removed or utilized [77]. In situations where serum lactate increases, there exists a fault along the metabolic pathway which could be on the oxygen side of the pyruvate-lactate pair, and hence in the oxidative metabolism, which takes place in the mitochondria [77].

In their classical work on energy expenditure of children with kwashiorkor, Ablett and McCane [78] showed that the BMR of kwashiorkor subjects were subnormal, and rose after a few days or weeks of treatment to the normal range. The low BMR persisted in untreated kwashiorkor. However, the best reference standard against which to express the metabolic rates of the malnourished infants has not been agreed on conclusively.

Keys et al. [79] in 'Biology of human starvation' felt that the cause of low BMR has not been elucidated in kwashiorkor. Brenton et al. [80] observed that hypothermia in kwashiorkor is associated with an increased mortality rate of the kwashiorkor children. Speculations regarding the low BMR in kwashiorkor include that of Waterlow [81] who felt that the metabolic pathways may be changed; Cohen and Hansen [82] and Hoffenberg et al. [83] postulated that low BMR may be the result of the low turnover rate of proteins in some organs during kwashiorkor. However, Ablett and McCane [78] asserted that the absence of all the normal metabolism of growth must certainly be a contributory factor in the declining energy expenditure of PEM patients.

It has been shown that the ability of the malnourished child to survive a continuing shortage of dietary energy and protein often in the face of increased energy requirements due to infection, is dependent on successful adaptation to various metabolic pathways [73]. Kerr et al. [84] have demonstrated that glucose production, due to gluconeogenesis when children were malnourished, was nearly the same as when they recovered. These workers, however, admitted that the exact mechanism of this adaptation needs elucidation. Efficient recycling of glycolytic products has been identified to be a contributory factor in the crucial production and bioavailability of this energy-giving molecule.

It has been shown that one of the striking features of kwashiorkor was described in 1947 in the original paper of Williams [85], namely: deposition of triglyceride in the liver has been shown to impair energy production in eukaryotic cells by uncoupling mitochondria. It has been shown that detoxication mechanisms, which invariably involve energy, are similarly impaired in malnourished subjects. The LD_{50} of diazinon, a pesticide, has been shown to be reduced in kwashiorkor rats [86]. Similarly, it has been shown that there is a defect in the ability of the marasmic-kwashiorkor rats to synthesize protein [87].

Table 4. Summary of the present series of experiments

	Average body weight	Predicted BMR	Measured RBMR	Percentage measured/ percentage predicted
Kwashiorkor	25.20 ± 4.58	1.53 ± 0.08	0.70 ± 0.08	45.75
Obese	115.00 ± 11.9	1.02 ± 0.04	1.07 ± 0.06	104.90
Control	71.20 ± 7.1	1.19 ± 0.04	0.87 ± 0.07	73.11

Age of rats at the end of experiment was 42 days. Feeding period was 21 days.

The incidence of severe hypoglycaemia which is accompanied by hypothermia has contributed to the high mortality rate in kwashiorkor children [73, 80]. Low BMR and reduced energy expenditure have also been observed in PEM [78]. The exact mechanism of these deranged bioenergetics during PEM has not been defined [88].

To unravel the mechanisms of the bioenergetics of PEM as well as those suffering from dietary obesity, the author (1976–1990) has painstakingly investigated the structural integrity and the energy-linked functions of hepatic mitochondria isolated from rats which are suffering from these malnutritional diseases. The synopsis of these experiments and their results (table 4) including closely related works from other laboratories are brought into focus in this review.

The metabolic aberrations during obesity outstrip that of PEM. The complexities of defective thermogenesis, hyperglycaemia and DIT are metabolic peculiarities in obesity. Figure 7 shows the numerous metabolic defects in obesity syndrome. It is the belief of this author that the bioenergetic problems in obesity far outstrip that in kwashiorkor and/or marasmus. The singular presence of aberrant behaviour found in the brown adipose tissue (BAT) mitochondria present in obese subjects supports this assertion [89]. Details of these aberrant bioenergetics in the face of surplus caloric intakes are focussed in the next two succeeding sections. The mitochondrion has been used as the tool of these reported results.

Uncoupling of Mitochondria by Exogenous and Endogenous Substances

A number of endogenous and exogenous compounds have been found to interact with the respiratory chain in such a way that the process of reduction of molecular oxygen to water is inhibited. Examples of such compounds are: (a) antimycin A [90, 91]; (b) Amytal [13, 92, 93]; (c) Rotenone [56, 94]; (d) cyanide [95, 96]; (e) oligomycin [91, 97–99]; (f) the dinitrophenols [100, 101]; (g) dicumarol [102]; (h) fluoro-chloro-carbonyl-phenylhydrazine [98]; (i) free fatty acids [13]; (j) N-phosphonomethylglycine (otherwise known as glyphosate [103]; (k) some naturally occurring benzoquinones (including the antifertility drug, Embelin) [104].

The interactions of uncouplers with mitochondrial membranes interfere with oxidative phosphorylation characteristics and hence ATP generation in the cell. Some of them (listed above) have been used as selective tools to study and elucidate the mitochondrial energy-transducing mechanisms.

Roles of Defective Energy Metabolism in the Aetiology of Obesity

Defective Thermogenesis in Obesity

It has been documented by Jequier and Shutz [4] that a thermogenic defect plays a role in the development of a number of genetically transmitted obesities. These subjects exhibit both hyperphagia and increased metabolic efficiency. It has been experimentally shown that during pair-feeding studies with lean controls, ob/ob mice became obese [105]. These authors inferred that the reduced thermogenesis alone can be responsible for obesity in these mutants. In addition, hypothalamic obesity, induced in the rat by lesions in the ventromedial hypothalamus (VHM), also results into hyperphagia and an increase in metabolic efficiency [106].

The concept of thermogenic defect in obese subjects has gained credence as a result of studies showing a reduced postprandial response in obese individuals after glucose [107, 108].

The obese subjects, in whom a thermogenic defect was observed, were usually selected on the basis of family history of obesity [109]. However, conflicting reports showing an unaltered thermogenic response to a meal have also been documented [110]. Dietary obese subjects, not selected on the basis of family history, have also been found to demonstrate thermogenic defects. The reasons for these contradictory results are: (a) the multi-aetiology of obesity syndrome as well as the extent of adipocity, peripherally or abdominally, and (b) the methodological problems, which vary from one investigator to another [4].

Techniques better than those currently being used might have to be employed to unequivocally resolve the contradiction. The various techniques being used include: (a) the use of respiratory chamber; (b) extrapolation from basal metabolic rate; (c) diet-induced thermogenesis (DIT); (d) extrapolation from the energy expenditure due to physical activity, and (e) the use of direct and indirect calorimetry.

The inherent limitations of the various methods vis-à-vis thermogenesis can be a factor in the contradictory results from various laboratories. Since there is a consensus that there exists a *thermic* effect of food, the relevant issue therefore to clarify is the significance of the reduced thermogenic response to food ingestion, as a factor, favouring the aetiology of obesity. The use of a single common technique for all forms of obesity, i.e. bioenergetic techniques, to ascertain the status of the energy transduction in mitochondria of obese animals is advocated and used. The results obtained from the laboratory of this author and other laboratories are hereby reported to provide an 'energy baseline' in obesity.

Changes in Resting Basal Metabolism in Dietary Obesity

In a series of experiments carried out by the author and his collaborators in his laboratory as well as in the Department of Biochemistry, University of Nairobi, Kenya, it was experimentally demonstrated that there were notable changes in the resting basal metabolism in dietary obese rats.

The findings above confirmed the earlier observation that diet-induced obese subjects have higher BMR (expressed in absolute terms) and a greater overall resting energy expenditure despite their reduced thermogenic response to food ingestion.

It has been reported that obesity is usually accompanied by an increased energy expenditure and a reduction in DIT [110] while kwashiorkor has been associated with hypothermia and defective ATP formation.

Between 1976 and 1990, this author systematically assessed the bioenergetics in malnutritional states. The summary of results show that there is an aberration in energy metabolism in both undernutrition culminating in kwashiorkor and overnutrition resulting in obesity [16]. This author also opines that these bioenergetic problems in obesity and kwashiorkor are basically secondary factors rather than primary. They are better described as one of the 'immediate consequences' of overnutrition and malnutrition in general. In spite of the divergent aetiology of obesity and kwashiorkor from the dietary viewpoint, both share in common some bioenergetic aberrations. This common denominator, reported in the next section, provides an experimental basis for the generalization [112] that 'nutritionally unbalanced diets invariably lead to defective energy metabolism'. It can be surmised that whilst these bioenergetic parameters are mostly of molecular and enzymatic origin, the only apparently superficial indicator, i.e. BMR or RBMR, could be measured using the method of Kleiber [113]. This, no doubt, can be of clinical application in the prognosis of obesity syndrome [28].

Changes in Oligomycin-Sensitive ATPase in Dietary Obesity

Oligomycin is a typical inhibitor of oxidative phosphorylation [114]. This compound becomes a useful tool of investigating energy metabolism as it confers preferential sensitivity on one of the component proteins that form the basal part of the ATPase complex. The oligomycin sensitivity-conferring protein thus gives an indicator of mitochondrial functionality, hence the index of the formation and hydrolysis of ATP to form both useful biological energy as well as heat. It was noted that ATPase activity in mitochondria from obese rats was less oligomycin sensitive (oligomycin sensitivity was 60–80%) than in the control animals where oligomycin sensitivity was between 97 and 99% [15]. This observation points to yet another inherent lesion at the mitochondrial level in dietary obese rats. It may also assist in understanding the rather complex bioenergetics of obese subjects.

Na^+-K^+-ATPase and Basal ATPase Status in Obesity Syndrome

The multi-subunit enzyme (ATPase) becomes very significant in the consideration of formation/hydrolysis of ATP because of the importance of the equation:

$$ATP \xrightleftharpoons[\text{ATPase}]{\text{ATPase}} ADP + Pi + \text{Energy useful} + \text{Heat.}$$

Studies on the mitochondrial ATPase measured in the direction of ATP hydrolysis indicate that both the basal and carbonylcyanide-m-chlorophenylhydrazone induced activity in obese rats was about 40% lower than in control. This observation suggests that in obese rats, the enzyme has lower capacity to hydrolyse ATP than in control rats, thus suggesting a lower rate of energy generation through ATP hydrolysis [15]. This may further explain the reduced thermogenesis observed in obesity [4, 29]. Table 5 gives a summary of results in mitochondria isolated from obese and control rats.

Table 5. ATPase activity in mitochondria isolated from obese and control rats [from 15]

Mitochondrial source	ATPase activity, nmol/min/mg protein			
	basal			CCCP-induced + oligomycin
	oligomycin	+ oligomycin	oligomycin	
Obese rats	215.00	78.00	480.00	96.00
Control rats	370.00	10.00	720.00	8.00

CCCP = carbonylcyanide-*m*-chlorophenylhydrazone.

The results from our laboratory in our studies on the status of both the basal and Na^+-K^+-ATPase are in agreement with those of Izpisua et al. [115]. Olowookere et al. [15] showed that the mitochondrial membranes in dietary obese rats show greater fluidity as well as significant ($p > 0.001$) decreases in Na^+-K^+-ATPase which is a membrane-bound enzyme just like most other enzymes of the inner mitochondrial membranes.

A possible conclusion from these observations are: (a) changes in the fluidity of obese rat liver mitochondria membranes; (b) decreases in basal ATPase measured along the direction of ATP hydrolysis; (c) decreases in the Na^+-K^+-ATPase, and (d) a possible 'partial uncoupling' of the mitochondrial membranes which consequently may disrupt the topology of the ATPase-enzyme complex. The multi-subunit nature of ATPase (basal and Na^+-K^+) makes it rather vulnerable to fluidity and hence deviation from vectorial chemiosmosis [72]. The changes in the cholesterol and phospholipid constituents of the obese rat mitochondrial membranes may contribute to the alteration in the membrane texture, membrane fluidity and consequently depresses the activity of the basal ATPase as well as Na^+-K^+-ATPase in dietary obesity.

The Energy Implications of Cyclical Obesity in Hibernators

The fall fattening and winter weight loss of mammalian hibernators has been known since antiquity [116]. There is now ample evidence to support the view that in hibernators just as in nonhibernators, ablation of the VMH and lateral hypothalamus (LH) resulted respectively in obesity and weight loss [117]. The golden-mantled ground squirrel *(Spermatophi-*

lus lateralis) has been used as a model to study cyclical obesity in mammalian hibernators. The evolution of thermoregulatory and endocrine cycles in hibernators has not been seriously and rigorously investigated. However, modern studies of regulatory physiology have made several unsuccessful efforts to identify the adjustable regulator for food intakes, as well as weight gains vis-à-vis the energy implications in hibernators. The number of mitochondria per cell during hibernation is reduced. This occurs as part of preparation towards hibernation. The energy requirements of hibernators being necessarily lower than nonhibernators, the number of mitochondria per cell is correspondingly reduced. The production of mitochondria as cellular organelles follows mitotic pattern; this means that it is controlled by the genome. The cyclic obesity in hibernators is probably under the control of the genome, which in turn is traceable to evolution. The 'intractable' regulator of obesity in hibernators may be a result of the complexity of mammalian genome. The answer may well lie in the modern techniques of genetic engineering. This author agrees with Mrosovsky [116] that whilst there were justifiable reasons, 20 years ago, to focus on the hypothalamus on which experiments on weight cycles revolve despite VMH and LH lesions, it is now timely to turn our attention elsewhere in the search for the adjustable regulator in cyclical obesity in hibernators. So far, no treatments are known to abolish weight cycles in hibernators. These cycles persist despite removal of the thyroids, gonads, pineal, olfactory bulb and a variety of hypothalamic and midbrain lesions. There is a hint that certain midbrain lesions attenuate or disrupt cyclicity of obesity in hibernators, but the present data are inconclusive [116].

Perhaps the search for the adjustable regulator of weights in hibernators should be a multidisciplinary one focussing on a *biochemical molecule* (e.g. a section of DNA, a protein or enzyme or an endogenous pharmacological determinant of behavioural and physiological changes). Using the dormouse (another mammalian hibernator model), it was found that there was notable insulin seasonal fluctuations [18]. It was also found that adipocytes had lower glucose-6-phosphate dehydrogenase activity and oxidized less glucose in response to insulin in the summer, but these indices changed in the fall and winter when the animals are in lipogenic state.

The above, namely poor oxidation of glucose during the summer relative to the winter seasons, provided the first comprehensive profile of carbohydrate metabolism in hibernators and demonstrated marked seasonal changes. It also provided a direct evidence of the lesions in mitochondria isolated from obese animals. More importantly, however, the

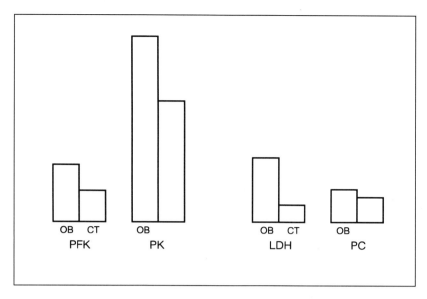

Fig. 8. Activity profiles of four glycolytic enzymes in control (CT) and dietary obese (OB) rats. PFK = Phosphofructokinase; PK = pyruvate kinase; LDH = lactic dehydrogenase; PC = pyruvate carboxylase.

findings provide a solid scientific basis to focus on the status of glycolytic energy production which is an obligatory precursor of mitochondrial phase of energy production in eukaryotic cells.

Changes in Key Glycolytic Enzymes in Dietary Obesity

Glycolysis, otherwise known as the Embden-Meyerhof pathway of sugars, yields pyruvic acid and lactic acid. Glycolysis in tissues consists of the breakdown of glycogen, glucose, or other sugars to pyruvic and lactic acids. It is a process of carbohydrate metabolism generally characteristic of animal cells. Although one stage of glycolysis requires oxidation by dehydrogenation, this is often accomplished without oxygen. Essentially the process of glycolysis is anaerobic. The glycolytic process is necessary for most phases of carbohydrate metabolism except the interconversion of sugars to glycogen. It is an obligatory pathway of carbohydrate oxidation since the pyruvic acid formed is oxidatively decarboxylated to form acetyl fragments (C_2). The cyclic processes represented in the tricarboxylic acid

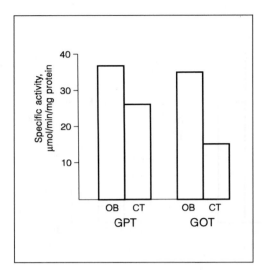

Fig. 9. Activity profiles ($\times 10^3$) for glutamate pyruvate transaminase (GTP) and glutamate oxaloacetate transaminase (GPT) in control (CT) and dietary obese (OB) rats.

(TCA) cycle takes on this two-carbon fragment via coenzyme A (i.e. acetyl-CoA) to form ATP, CO_2, and H_2O. The TCA processes are localized in the mitochondria.

Kerr et al. [88] have shown that there is a rapid recycling of glycolytic intermediates in protein-energy malnutrition. Mela et al. [77] have established that whenever there is a fault in the mitochondrial energy transduction, there may likely be a corresponding compensatory faster recycling of glycolyis. It is with this background and with a view to clearly unravel the bioenergetics of obese subjects that prompted the essay of key glycolytic enzymes from the liver homogenate of animals suffering from dietary obesity.

The activity profiles of four key glycolytic enzymes assayed from the liver homogenate of animals suffering from dietary obesity were investigated in our laboratory [17]. These enzymes are phosphofructokinase, pyruvate kinase, LDH, and pyruvate carboxylase. Figure 8 shows the activity profiles of four glycolytic enzymes in control and dietary obese rats. In addition, the activities of the cytosolic glutamate pyruvate transaminase and glutamate oxaloacetate transaminase were determined (fig. 9) [17].

It has been documented that there is an increase in LDH in under-nourished (kwashiorkor) patients [75]. Elevation of lactate, the end-product of anaerobic respiration, suggests that glucose metabolism passes freely through the glycolytic Embden-Meyerhof pathway to lactate. The logical inference from this aberration was succinctly put by Mela et al. [77] that 'there exists a fault along the metabolic pathway which could be on the oxygen side of pyruvate-lactate pair and hence in the oxidative metabolism which takes place in the mitochondria'. The various increases in the glycolytic enzymes, i.e. phosphofructokinase, pyruvate kinase, LDH and pyruvate carboxylase, further buttress the inference that 'some faults' may exist in the mitochondria isolated from obese rats.

Limitations of Direct and Indirect Calorimetry in the Assessment of Bioenergetics in Dietary Obesity

Calorimetry is the measurement of energy expenditure either as heat loss (direct calorimetry) or as respiratory gas exchange to determine metabolic rate (indirect calorimetry). Calorimetric measurements are often made in the context of dietary control for some extended period. Usually it is accompanied with measures of body composition to assess energy stores. Essentially, calorimetry determines the 'significant features' in energy balance [12].

Webb [12] documented that obese and overweight people do not differ from others in obvious ways. He asserted that fat people are not more efficient in handling food nor are lean people extravagant 'spenders' of energy, at least under controlled conditions. While Webb [12] agreed that there is an increased thermogenesis during overeating and decreased thermogenesis during undereating, he concluded by stating that differences in their responses to dietary intake are not clear from experimental data. Jung et al. [120] proposed that the mechanism of the increased thermogenesis may be related to response to catecholamines. These workers infused noradrenaline into lean and obese subjects, which caused a sizeable increase in metabolism in the lean subjects but only half that response in the obese subjects.

Webb [12], having observed some of the inherent limitations in the use of direct and indirect calorimetry, suggested that future research could employ better techniques for monitoring activity vis-à-vis food intake. The same author opined that such new techniques should make it possible to learn how people regulate energy stores, or fail to, in the face of ordinary stresses such as anxiety, temperature changes and variations in food intake. He concluded that such futuristic techniques may enable researchers in this field to understand how energy balance is controlled in the obesity syndrome.

An attempt to pave the way to the evolution of better clinical techniques, such that researchers and clinicians could understand how energy balance is controlled in dietary obesity, has led to a series of studies on the bioenergetics of obese animal models. The mitochondrion which is the power plant of the cell has been focussed upon and used as a model. The various observations at the anatomical level resulting in pathological conditions are all known to emanate from lesions at the molecular level of cellular organization. Hence, the succeeding section summarizes the various lesions and aberrations found in mitochondria isolated from dietary obese animal models.

Mitochondrial Status and Related Energy-Linked Functions of Mitochondria Isolated from Normal and Dietary Obese Animals

Changes in Phospholipids, Cholesterol and Total Lipids in Undernourished and Overnourished (Obese) Animals

It has been shown that in mitochondria isolated from undernourished (kwashiorkor) rats, significant decreases occurred in the concentration of total lipids and phospholipids, while significant increases occurred in the total cholesterol and triglycerides in control pair-fed rats. In obese rats the values of total lipids and cholesterol exceeds that of the values in control (normal) rats. The values for total lipids and cholesterol are of the order kwashiorkor < control < obese.

The phospholipid values in the deficient (kwashiorkor) group mitochondria were 4.8–5.1 mg phosphorus/g protein, and 8.6–10.7 mg/g protein in obese rats in comparison with 5.7–6.6 mg/g protein in the controls. Cholesterol levels were 6.35 mg/g protein in deficient rats and 4.6 mg/g protein in controls. However, in obese animals the value of cholesterol was 9.2 mg/g protein. The increase in total cholesterol in mitochondria from the deficient rats is consistent with previous observations that total liver cholesterol was increased in rats suffering from vitamin A deficiency [121]. It has also been shown that rats suffering from protein-energy malnutrition lacked the ability to utilize vitamin A and hence behave essentially like rats suffering from vitamin A deficiency [122]. Changes in the levels of cholesterol and phospholipids in mitochondria from deficient rats suggest that vitamin A may be an integral structural component of the mitochondrial membrane [122]. However, the increases in the mitochondria total lipid and cholesterol may alter the critical lipid/protein ratio of mitochondrial membrane.

Changes in mitochondrial membrane texture occur when there are changes in the constituent lipids, proteins and other smaller molecules like vitamin A and manosyl fragments [123, 124]. Izpisua et al. [115] reported that mitochondrial membranes isolated from dietary obese rats exhibited greater fluidity than normal rat liver mitochondrial membrane. It is therefore logical to speculate at this stage that mitochondria from dietary obese rats may also not function optimally.

Deviation from Vectorial Chemiosmosis and Changes in Oxidative Phosphorylation Characteristics of Obese Animals

Mitchell [72] in his 'vectorial chemiosmosis' clearly demonstrated that the direction and magnitude of flow of electrons in normal mitochondria followed a simple vector. This theory, i.e. chemiosmotic theory, was experimentally demonstrated by the same author in 1979 [125]. Olowookere [62, 124] demonstrated that there is a deviation from vectorial metabolism in the hepatic mitochondria of PEM rats.

Recently, Izpisua et al. [115] demonstrated that mitochondrial membranes isolated from dietary obese rats show greater fluidity – a situation that points to possible deviation in vectorial metabolism of obese rat liver mitochondria. Olowookere et al. [15] experimentally showed that there are significant alterations in the oxidative phosphorylation characteristics of dietary obese rat liver mitochondria. These authors [15] showed that the resting metabolic rate in obese animals was 23% higher than the theoretically predicted value. Although the mitochondrial oxygen consumption pattern, using malate plus glutamate or succinate as respiratory substrate, revealed that the resting respiration (state 4) was 29.1% higher in obese rats, the active state 3 (ADP-dependent) respiration was 43.3% lower in obese rats compared to controls. The respiratory control ratio (RCR), which is the biochemical index for ATP synthesis, was 43.3% in obese rat liver mitochondria relative to control. From these studies it was concluded: (a) that dietary obese animals' mitochondria have reduced RCR, and (b) dietary obesity interferes with mitochondrial oxygen utilization at the level of state 3 (ADP-dependent) mitochondrial respiration. In summary, there was an adjustment upwards in state 4 mitochondrial respiration while there was a drastic reduction in the state 3 (ADP-dependent) active respiration of mitochondria isolated from dietary obese rats.

Since each of the electron complexes specifically functions for each of the three sites of the mitochondrial oxidative phosphorylation in a mechanistic manner that is vectorial, it follows that there is an attendant alteration in the electron transfer complexes during obesity. This view is consistent with the abnormal fluidity observed by Izpisua et al. [15] and the reduced RCR and ADP/O ratio reported by Olowookere et al. [15].

ATPase Status and Brown Adipose Tissue (BAT) Mitochondria Isolated from Obese Animals

There are a number of energy-requiring metabolic cycles in the body which participate in the expenditure of energy under basal as well as diet-induced conditions. These include the hydrolysis of ATP to ADP + Pi by Na^+-K^+-ATPase [89, 126] as well as the so-called 'futile cycles' involved in the bioenergetics of obese subjects. Izpisua et al. [115] have shown that there is a decrease in Na^+-K^+-ATPase of mitochondria isolated from dietary obese rats.

BAT mitochondria contain yet another pathway for the release of energy as heat. This proton conductance pathway allows for the leakage of protons into the mitochondria when nucleotides are loosely bound to a 32,000 molecular weight protein on the inner mitochondrial membrane [127]. Under these conditions, free fatty acids released during intracellular lipolysis are oxidized within mitochondria with the subsequent release of heat, rather than formation of ATP or any of the high energy phosphate bonds [89]. As such, BAT represents a major source of nonshivering thermogenesis in the rat [128] and may as well be a contributory factor in DIT. Regardless of the primary source of thermogenesis, the sympathetic nervous system appears to be the most important effector of this process in mammals [129].

Respiratory Impairments and Low Work Capacity in Obese Subjects

The respiratory impairment and defects in mitochondrial respiration reported by Olowookere et al. [15] brought to focus eight bioenergetic problems, namely: (a) defective ATP synthesis; (b) defective ATP hydrolysis; (c) poor utilization of molecular oxygen; (d) deviation from vectorial metabolism resulting from abnormal fluidity of mitochondrial membranes; (e) reduced RCR; (f) reduced ADP/O ratio, and (g) abnormal mitochondrial guanosine diphosphate binding [130].

The resultant effects of all these biochemical lesions culminate in the observed low work capacity of obese subjects.

Does a Dietary Obese Animal Rely More on Glycolysis?

Compared to normal animals and the data presented in this review on the activity profiles of four key glycolytic enzymes (see section 'Changes in Key Glycolytic Enzymes in Dietary Obesity'), there is evidence to show that there may be greater reliance on glycolysis by obese animals than normal. The various mitochondrial lesions listed in the above section buttress this conclusion since the mitochondria isolated from the obese animals do not function optimally. The reliance on glycolytic energy production might be an adaptation or a compensatory mechanism, bioenergetically. This inference is consistent with the view of Kerr et al. [84] who observed a rapid recycling of glycolysis and an increase in gluconeogenesis in malnourished children.

Conclusion and Future Directions

This review on the bioenergetics of obesity syndrome represents a condensation of facts, ideas and laboratory findings on the role of energy metabolism in obesity syndrome. It has been able to highlight and synthesize the various data in the usually neglected and abstract field of bioenergetics in the last decade (1981–1991). The various findings reported would be of immense value to clinicians, nutritionists, pharmacologists, physiologist, community health workers and other paramedicals involved in the management and rehabilitation of obese subjects. It will definitely come handy to researchers in this field since such a piercing review on obesity bioenergetics is very scarce to come by, if at all available.

The pharmacologists, biochemists, dieticians and food policy-makers in the industrialized world, where obesity is most prevalent, would have in this review a handy bibliography for the energy parameters in obesity. These data vis-à-vis gross national product(s) and life expectancy of obese subjects may accelerate the evolution of better management techniques and strategies in the rehabilitation of subjects suffering from obesity. It is pertinent to mention that one striking revelation of this review is that obese subjects share some curious bioenergetic features in common with malnourished subjects [16].

Future research should be aimed at the assessment of the influence of the sympathetic nervous system and the biochemical roles of insulin on DIT in lean, obese and 'post-obese' individuals. A multidisciplinary approach towards the realization of the above objectives is advocated.

With these few suggestions, this review has opened up yet another 'pandoras box' for researchers in the areas of membrane biochemistry, clinical nutrition, pharmacology, physiology (including exercise physiology), dietetics, endocrinology, human/animal genetics vis-à-vis the obesity syndrome and its attendant myriad of health problems.

References

1　Mayer J: Obesity syndrome in man; in Goodhart RS, Shils ME (eds): Modern Nutrition in Health and Disease. Philadelphia, Lea & Febiger, 1978, pp 721–740.

2　Goodhart RS, Shils ME: Respiratory difficulties in obesity; in Goodhart RS, Shils ME (eds): Modern Nutrition in Health and Disease. Philadelphia, Lea & Febiger, 1978, pp 570–589.

3　Thurnly PL, Trayhurn P: The role of thermoregulatory thermogenesis in the development of obesity in genetically obese (ob/ob) mice pair-fed with lean siblings. Br J Nutr 1979;42:377–385.

4　Jequier E, Shutz Y: Does a defect in energy metabolism contribute to human obesity? In Hirsch J, Van-Itallie TP (eds): Recent Advances in Obesity Research. London, Libbey, 1983, pp 69–89.

5　Beaton GH: Nutritional problems of affluence; in Beaton GH, Bengoa JM (eds): Nutrition in Preventive Medicine. Geneva, WHO, 1976, pp 482–499.

6　Blackburn L, Greenberg G: Multidisciplinary approach to adult obesity therapy; in Hirsch J, Van-Itallie TP (eds): Recent Advances in Obesity Research. London, Libbey, 1983, pp 129–145.

7　Bandura A, Simon KM: The role of proximal intentions in self-regulations or refractory behaviour. Cogn Ther Res 1977;11:1917–1937.

8　Lew EA, Garfunkel L: Variations in mortality by weight among 750,000 men and women. J Chron Dis 1979;32:563–576.

9　Drenick EJ, Bale GS, Selzer F, et al: Excessive mortality and causes of death in morbidly obese men. J Am Med Assoc 1980;243:443–445.

10　Garrow JS: Energy balance and obesity in man. Amsterdam, Elsevier, 1978, pp 1–243.

11　Garrow JS: Treat obesity seriously: A clinical manual. Edinburgh, Churchill Livingstone, 1981, pp 1–129.

12　Webb P: Use of direct and indirect calorimetry in studying obesity: Current summary and directions for the future; in Hirsch J, Van-Itallie TP (eds): Recent Advances in Obesity Research. London, Libbey, 1983, pp 93–100.

13　Lehninger AL: The mitochondrion. New York, Benjamin, 1964, pp 1–263.

14　Olowookere JO: The bioenergetics of protein-energy malnutrition syndrome. World Rev Nutr Diet. Basel, Karger, 1987, vol 54, pp 1–25.

15　Olowookere JO, Konji VN, Omwandho CA, et al: Changes in oxidative phosphorylation characteristic in dietary obese rats. Biosci Res Commun 1991;3:219–226.

16　Olowookere JO, Konji VN, Makawiti DW, et al: Defects in resting metabolic rates and mitochondrial respiration in kwashiorkor and dietary obese rats. J Comp Physiol [B] 1991;161:319–322.

17 Olowookere JO, Konji VN, Makawiti DW, et al: Changes in four key glycolytic enzymes and cytosolic transaminases in rats suffering from dietary obesity. Int J Obes, submitted.

18 Greenwood MRC: Normal and abnormal growth and maintenance of adipose tissue; in Hirsch J, Van-Itallie TP (eds): Recent Advances in obesity Research. London, Libbey, 1983, pp 37–45.

19 Lavau M, Brazin R, Guerre-Millo M: Fatty acid synthesis capacity and activity during early development in lean and obese Zucker rats. Int J Obes 1984;8:96–102.

20 Gruental RG, Hietanen E, Greenwood MRC: Increased adipose tissue lipoprotein lipase activity during development of the genetically obese rat (fafa). Metabolism 1978;27:1966.

21 Turkenkopf IJ, Chow G, Goldstein AL, et al: Fatty acid (fa) gene effects in utero in Zucker rats. Fed Proc 1982;41:452–460.

22 Callaway CW, Greenwood MRC: Methods for characterizing human obesities: a progress report; in Hirsch J, Van-Itallie TP (eds): Recent Advances in Obesity Research. London, Libbey, 1983, pp 138–143.

23 Bray GA: Hypothalamic and genetic obesity: An appraisal of the autonomic hypothesis and the endocrine hypothesis. International Symposium on Novel Approaches and Drugs for Obesity. Int J Obes 1984;8:63–69.

24 DHEW Publ No (NIH) 76-852: Obesity in perspective, parts 1 + 2; in Bray GA (ed) Fogarty International Center Series on Preventive Medicine. Washington, US Government Printing Office, 1976, vol 2, pp 1–72.

25 Bray GA, Jordan HA, Sims EAH: Evaluation of the obese patient. I. An algorithm. J Am Med Assoc 1976;23:1487–1491.

26 Bray GA: Definition measurement and classification of the syndromes of obesity. Int J Obes 1978;2:99–112.

27 Sims EAH: Characterization of the syndromes of obesity; in Bleicher SJ, Brodoff BN (eds): Diabetes mellitus and Obesity. Baltimore, Williams & Wilkins, 1982, pp 219–226.

28 Olowookere JO, Olorunsogo OO, Malomo SO: Effects of defective in-vivo synthesis of mitochondrial proteins on cellular biochemistry of malnourished rats. Ann Nutr Metab 1990;34:147–154.

29 Sjorstrom L: Fat cells and body weight; in Stunkard AJ (ed): Obesity. Philadelphia, Saunders, 1980, pp 72–100.

30 Harrison GG: Anthropometric differences and their clinical correlates; in Hirsch J, Van-Itallie TP (eds): Recent Advances in Obesity Research. London, Libbey, 1983, pp 155–162.

31 Stunkard AJ: 'Restrained eating': What it is and a new scale to measure it; in Hirsch J, Van-Itallie TP (eds): Recent Advances in Obesity Research. London, Libbey, 1983, pp 243–251.

32 Smith U: Adrenergic control of human adipose tissue lipolysis. Eur J Clin Invest 1980;10:343–344.

33a Steingrimsdottir L, Brasel JA, Greenwood MRC: Diet, pregnancy and lactation:

effects on adipose tissue, lipoprotein lipase and fat cell size. Metabolism 1980;29: 837–841.

33b Sjorstrom L, Smith U, Krotkiewski M, et al: Cellularity in different regions of adipose tissue in young men and women. Metabolism 1972;21:1143–1153.

34 Frankle TR: Dietotherapy in the treatment of disease; in Hirsch J, Van-Itallie TP (eds): Recent Advances in Obesity Research. London, Libbey, 1983, pp 176–178.

35 Sullivan AN, Triscari J: Pharmacologic approaches to the regulation of metabolism and obesity; in Hirsch J, Van-Itallie TP (eds): Recent Advances in Obesity Research. London, Libbey, 1983, pp 196–207.

36 Garattini S: Central mechanisms of anoretic drugs; in Hirsch J, Van-Itallie TP (eds): Recent Advances in Obesity Research. London, Libbey, 1983, pp 208–216.

37 Levin BE: Thermogenic agents as a treatment; in Hirsch J, Van-Itallie TP (eds): Recent Advances in Obesity Research. London, Libbey, 1983, pp 217–233.

38 Kral JG: Surgical approach to the treatment of obesity: Introduction; in Hirsch J, Van-Itallie TP (eds): Recent Advances in Obesity Research. London, Libbey, 1983, pp 234–236.

39 Wilson GT: Psychological prognostic factors in the treatment of obesity; in Hirsch J, Van-Itallie TP (eds): Recent Advances in Obesity Research. London, Libbey, 1983, pp 301–311.

40 Berchtold P, Van-Itallie TB: Physiological prognostic factors for the treatment of obesity; in Hirsch J, Van-Itallie TP (eds): Recent Advances in Obesity Research. London, Libbey, 1983, pp 320–326.

41 Himms-Hagen J: Role of the adrenal medulla in adaptation to cold: A Handbook of Physiol Endocrinol Adrenal Gland. Am Physiol Soc, Sect 7, 1975;6:637–665.

42 Steingrimsdottir L, Brasel JA, Greenwood MRC: Diet, pregnancy and lactation: effects on adipose tissue, lipoprotein lipase and fat cell size. Metabolism 1980;29: 837–841.

43 Frisancho AR, Flegel PN: Advanced maturation associated with centripetal fat pattern. Hum Biol 1982;54:717–727.

44 Stallones L, Mueller WH, Christensen BL: Blood pressure, fatness and fat patterning among USA adolescents from two ethnic groups. Hypertension 1982;4:483–486.

45 Herman CP, Polivy J: Restrained eating; in Stunkard AJ (ed): Obesity. Philadelphia, Saunders, 1980, pp 61–67.

46 Keesey RE: A set point analysis of the regulation of body weight; in Stunkard AJ (ed): Obesity. Philadelphia, Saunders, 1980, pp 23–32.

47 Brownell KD: Behavioural, psychological and environmental predictors of obesity and success at weight reductions. Int J Obes 1984;8:18–24.

48 Cummings C, Gordon JR, Marlatt GA: Relapse: Prevention and prediction; in Miller WR (ed): The addictive Behaviours: Treatment of Alcoholism, Drug Abuse, Smoking and Obesity. New York, Pergamon Press, 1980, pp 71–77.

49 Thompson JK, Jarvie GJ, Lahey BB, et al: Exercise and obesity: aetiology, physiology and interventions. Psychol Bull 1982;91:55–79.

50 Brownell KD, Stunkard AJ: Exercise in the development and control of obesity; in Stunkard AJ (ed): Obesity. Philadelphia, Saunders, 1980, pp 14–29.

51 Cobb S: Social support as a moderator of life stress. Psychosom Med 1976;38:300–314.

52 Garn SM, Clark DC: Trends in fatness and the origins of obesity. Pediatrics 1976; 57:433–456.

53 Zuti WB, Golding LA: Comparing diet and exercise as weight reduction tools. Physician Sports Med 1976;4:49–57.

54 Xarier Pi-Sunyer: Effect of exercise on food intake; in Hirsch J, Van-Itallie TP (eds): Recent Advances in Obesity Research. London, Libbey, 1983, pp 312–324.

55 Lowenstein JM: Tricarboxylic acid cycle; in Greenberg DM (ed): Metabolic Pathways. New York, Academic Press, 1976, pp 146–270.

56 Ernster L: The phosphorylation occurring in the flavoprotein region of the respiratory chain; in Slater EC (ed): Symposium on Intracellular Respiration: Phosphorylating Oxidation Reactions. Proc 5th Int Congr Biochem, Moscow 1961. Oxford, Pergamon Press, 1963, pp 115–156.

57 Boyer PD, Chance B, Ernster L, et al: Oxidative phosphorylation and photo-phosphorylation. Annu Rev Biochem 1977;46:995–1025.

58 Tzagologt A, Rubin MS, Sierra MF: Biosynthesis of mitochondrial enzymes. Biochim Biophys Acta 1973;301:71–104.

59 Adenuga GA, Mazaev AG, Olowookere JO: The peculiarities of proteins and RNA syntheses in the hepatic mitochondria of young and old animals: Studies without the use of an antibiotic. Exp Gerontol 1988;23:35–41.

60 Olowookere JO, Babubunmi EA, Bassir O: Effect of dietary protein intake on mitochondrial oxidative phosphorylation in the rat. Niger J Nutr Sci 1980;1:69.

61 Olowookere JO, Olorunsogo OO, Bababunmi EA: Oxidative phosphorylation characteristics of mitochondria isolated from marasmic-kwashiorkor rats. Proc 12th Int Congr Nutr, San Diego, Calif 1981, pp 1–71.

62 Olowookere JO, Olorunsogo OO: Effects of dietary protein deprivation on electron transfer complexes in hepatic mitochondria of weanling rats. J Anim Physiol Anim Nutr 1985;54:1–6.

63 Olowookere JO: Cytochrome oxidase status in protein-energy deficient rats. Ann Nutr Metab 1986;30:47–53.

64 Ku HS, Pratt HK, Spurr AR, et al: Isolation of active mitochondria from tomato fruit. Plant Physiol 1968;43:883–887.

65 Click RE, Hackett DP: The role of protein and nucleic acid synthesis in the development of respiration in potato tuber slices. Proc Natl Acad Sci USA 1963;50:243–250.

66 Nadakavukaren MJ: Fine structure of negatively stained plant mitochondria. J Cell Biol 1964;23:193–195.

67 Parsons DF, Bonner WD Jr, Verboon VC: Electron microscopy of isolated plant mitochondria and plastids using both the thin section and negative staining techniques. Can J Bot 1965;43:647–655.

68 Baker JE, Elfvin LG, Biale JB, et al: Studies on ultrastructure and purification of isolated plant mitochondria. Plant Physiol 1968;43:2001–2022.

69 Ikuma H: Electron transport in plant respiration. Annu Rev Plant Physiol 1972;23:419–436.

70 Olowookere JO: Phase-contrast molecular model for biological membranes (short report); in Kon et al (eds): Contemporary Issues in Biochemistry: Proc 4th FAOB Congr, Singapore 1986, ICSU Short Rep, vol 6, pp 1–16.

71 Mitchell P: The chemical and electrical components of the electro-chemical potential of H-ions across the mitochondrial cristae membrane. FEBS Symp 1969;17:219–232.

72 Mitchell P: Vectorial chemiosmosis. Annu Rev Biochem 1977;46:996–1005.

73 Alleyne GAO, Hay RW, Picou DI, et al: Protein-energy malnutrition. London, Arnold, 1977, pp 1–234.

74 Trowell HC, Davies TNP, Dean RFA: Kwashiorkor. London, Arnold, 1954, pp 1–49.

75 Baig HA, Edozien JC: Carbohydrate metabolism in kwashiorkor. Lancet 1965;7414: 661–665.

76 Standstead HH, Safwat SA, Ananda FS, et al: Kwashiorkor in Egypt. I. Clinical and biochemical studies with special references to plasma zinc and serum lactic dehydrogenase. Am J Clin Nutr 1965;17:15–26.

77 Mela L, Bacalzo LV Jr, Miller LD: Defective oxidative metabolism of rat liver mitochondria in hemorrhagic and endotoxin shock. Am J Physiol 1971;220:571–577.

78 Ablett JC, McCane RA: Energy expenditure of children with kwashiorkor. Lancet 1971;i:517–519.

79 Keys A, Brozek J, Henschel A, et al: The Biology of Human Starvation. Minneapolis, University of Minnesota/Oxford, Oxford University Press, 1950, pp 1–497.

80 Brenton DP, Brow RE, Wharton R: Hypothermia in kwashiorkor. Lancet 1967; 7304:416–416.

81 Waterlow JC: Observation of the mechanisms of adaptation to low-protein intakes. Lancet 1981;ii:1091.

82 Cohen S, Hansen JDL: Metabolism of albumin and globulin in kwashiorkor. Clin Sci 1962;231:351–359.

83 Hoffenberg R, Black E, Brock JF: Albumin and tracer studies in protein depletion states. J Clin Invest 1966;45:143–152.

84 Kerr DS, Stevens CG, Robinson HM, et al: On gluconeogenesis in the malnourished child; in Alleyne GAO, Hay RW, Picou DI, et al (eds): Protein-Energy Malnutrition. London, Arnold, 1977, pp 62–65.

85 Williams RT: Detoxication Mechanisms. London, Chapman & Hall, 1947, pp 1–490.

86 Boyd EM, Carsky E: Kwashiorkor genic diet and diazinon toxicity. Acta Pharmacol Toxicol 1969;27:284–294.

87 Enwonwu CO, Stambaugh RA, Sereebny L: Synthesis and degradation of liver ribosomal RNA in fed and fasted rats. J Nutr 1971;101:337–345.

88 Kerr DS, Stevens MG, Picou D: Proc 2nd Int Conf Stable Isotopes. National Technical Information Services. Springfield, US Department of Commerce, 1976, pp 17–21.

89 Horwitz BA: Cellular events underlying catecholamine-induced thermogenesis: Cation transport in brown adipocytes. Fed Proc 1979;38:2170–2176.

90 Ahmad K, Schneider HG, Strong FM: Studies on the biological action of antimycin A. Arch Biochem Biophys 1950;28:281–294.

91 Slater EC: Mechanism of uncoupling of oxidative phosphorylation by nitrophenols in intracellular respiration. Phosphorylating and non-phosphorylating oxidation reactions. Proc 5th Int Congr Biochem, Moscow 1961, pp 352–364.

92 Jalling O, Lindberg O, Ernster L: On the effect of substituted barbiturates on mitochondrial respiration. Acta Chem Scand 1955;9:198–199.

93 Chance B, Ernster L, Garland PB, et al: Flavoproteins of the mitochondrial respiratory chain. Proc Natl Acad Sci USA 1967;57:1498–1505.

94 Burgos J, Redrearn ER: The inhibition of mitochondrial reduced nicotinamide-adenine dinucleotide oxidation by rotenoids. Biochem Biophys Acta 1965;110:475–483.

95 Slater EC: The constitution of the respiration chain in animal tissues. Adv Enzymol 1958;20:147–199.

96 Slater EC: The components of the dihydrocozymase oxidase system. Biochem J 1950;46:484–503.

97 Lardy HA, McMurray WC: Antibiotics as tools for metabolic studies: A survey of toxic antibiotics in respiratory and glycolytic systems. Arch Biochem Biophys 1958; 78:557–596.

98 Ernster L: Mitochondrial bioenergetics: Facts and ideas. FEBS Symp 1975;35:257–285.

99 Slater EC: Mechanism of uncoupling of oxidative phosphorylation by nitrophenols in intracellular respiration. Phosphorylating and non-phosphorylating oxidation reaction. Proc 5th Int Congr Biochem, Moscow 1961, pp 352–364.

100 Hemker HC: Lipid solubility as factors influencing the activity of uncoupling phenols. Biochim Biophys Acta 1962;63:46–54.

101 Parker VH: Uncouplers of rat liver mitochondrial oxidative phosphorylation. Biochem J 1965;97:658–663.

102 Chappell JB: Mitochondria and 2,4-dinitrophenol. Biochem J 1964;90:237–248.

103 Olorunsogo OO: Acute toxicity studies with glyphosate in the rat and mouse; unpubl PhD thesis, University of Ibadan, Nigeria 1976.

104 Makawiti DW, Konji VN, Olowookere JO: Interaction of benzoquinones with mitochondria interferes with oxidative phosphorylation characteristics. FEBS Lett 1990; 266:26–28.

105 Hallgren P, Raddatz E, Bergh CH, et al: Oxygen consumption in collagenase liberates rat adipocytes in relation to cell size and age. Metabolism 1984;33:897–903.

106 Garrow JS: Energy Balance and Obesity in Man. New York, Elsevier, 1974, pp 1–80.

107 Golay A, Schutz Y, Meyer HU, et al: Glucose-induced thermogenesis in non-diabetic and diabetic obese subjects. Diabetes 1982;31:1023–1037.

108 Pittet PH, Chappuis PH, Achesin K, et al: Thermic effect of glucose in obese subjects studied by direct and indirect calorimetry. Br J Clin Nutr 1976;35:281–292.

109 Bessard T, Schutz Y, Jequier E: Energy expenditure and postprandial thermogenesis in obese women before and after weight loss. Am J Clin Nutr 1983;38:680–693.

110 Felig P, Cunningham J, Lenitt M, et al: Energy expenditure in obesity in fasting and postprandial state. Am J Physiol 1983;244:E45–E51.

111 James WPT, Bailes J, Davies HL, et al: Elevated metabolic rates in obesity. Lancet 1978;i:122–125.

112 Kleiber M: Dietary deficiencies and energy metabolism. Nutr Abstr Rev 1945;15:207–220.

113 Kleiber M: The Fire of Life: An Introduction to Animal Bioenergetics. New York, Wiley, 1961, pp 1–30.

114 Lardy HA, Cornelly JL: The site of action of uncoupling and specific inhibitors of oxidative phosphorylation; in Slater EC (ed): Symposium on Intracellular Respiration: Phosphorylating Oxidation Reactions. Proc 5th Int Congr Biochem, Moscow 1963. Oxford, Pergamon press, 1963, pp 365–400.

115 Izpisua JC, Barber T, Cobo J, et al: Lipid composition, fluidity and enzymatic activities of rat liver plasma and mitochondrial membranes in dietary obese rats. Int J Obes 1989;13:513–542.

116 Mrosovsky N: Cyclical obesity on hibernators: the search for adjustable regulator; in Hirsch J, Van-Itallie TB (eds): Recent Advances in Obesity Research. London, Libbey, 1983, pp 61–68.

117 Satinoff E: Aberration of regulation in ground squirrels following hypothalamic lesions. Am J Physiol 1967;212:1215–1220.

118 Vilberg TR, Beaty WW: Behavioural changes following VMH lesions in rats with controlled insulin level. Pharmacol Biochem Behav 1973;3:377–384.

119 Webb P: Energy expenditure and fat-free mass in men and women. Am J Clin Nutr 1981;34:1816–1826.

120 Jung RT, Shetty PS, James WPT: Reduced thermogenesis in obesity. Nature 1979; 279:322–323.

121 Mclaren DS, Tchalian M, Ajans ZA: Biochemical and haematologic changes in vitamin A-deficient rat. Am J Clin Nutr 1965;17:131–138.

122 Olowookere JO: Consequences of defective vitamin A transportation on mitochondrial integrity. Ann Nutr Metab 1986;30:210–212.

123 Olowookere JO: Biomembranes, vitamin A and the kwashiorkor syndrome. Proc 2nd Afr Nutr Congr, Ibadan 1983, p 17.

124 Olowookere JO: The fluid-mosaic model of biomembranes: Roles of exogenous dietary intake on its dynamic and functionality. Proc 17th Annu Conf of Nutr Soc of Nigeria, 1984, pp 4–5.

125 Mitchell P: Direct chemiosmotic ligand conduction mechanisms in proton motive complexes. Abstr Int Workshop on Membrane Bioenergetics, Detroit, Mich 1979, pp 1–51.

126 Rothwell NJ, Stock MJ, Wyllie MG: Na^+,K^+-ATPase activity and noradrenaline turnover in brown adipose tissue of rats exhibiting diet-induced thermogenesis. Biochem Pharmacol 1981;30:1702–1709.

127 Nicholls DG: Brown adipose tissue mitochondria. Biochem Biophys Acta 1979;549: 1–29.

128 Foster DO, Frydman ML: Non-shivering thermogenesis in rat. II. Measurement of blood flow with microsphere point to brown adipose tissue as the dominant calorigenesis induced by noradrenalin. Can J Physiol Pharmacol 1978;56:110–112.

129 Maickel RP, Matussek N, Stern DN, et al: The sympathetic nervous system as a homeostatic mechanism. I. Absolute need for sympathetic nervous functions in body temperature regulation in cold-exposed animals; in Hirsch J, Van-Itallie TB (eds): Recent Advances in Obesity Research. London, Libbey, 1983, pp 184–190.

130 Zanko MT, Sulliran AC, O'Brien RA: Defective purine nucleotide binding in brown adipose tissue mitochondria of genetically obese rats. Fed Proc 1982;41:714–719.

Julius O. Olowookere, Biomembranes and Bioenergetics Research Laboratory, Department of Biochemistry, Ogun State University, PMB 2002, Ago-Iwoye (Nigeria)

Subject Index